The American AZ Hive

By Debra Langley-Boyer

Northern Bee Books

The American AZ Hive

Copyright © Debra Langley-Boyer

Published 2025 by Northern Bee Books,
Scout Bottom Farm,
Mytholmroyd,
West Yorkshire
HX7 5JS (UK)
Tel: 01422 882751
Fax: 01422 886157
www.northernbeebooks.co.uk

ISBN 978-1-9192004-1-5

Design and artwork DM Design and Print

Bee

Table of Contents

Introduction, Dedication and Thank you

Chapter 1 : History

Chapter 2 : Getting started

Chapter 3 : The Hive

Chapter 4 : The bee house

Chapter 5 : Working the AZ Hive

Chapter 6 : Designs for the AZ

References & Credits

Preface

The journey to creating the *American AZ Hive* started in 2016 when my daughter, Melissa Baltzell, said, "mom let's take a beekeeping class and get bees together." So we did, and now I have bees, but she does not. Thank you, Melissa, for helping your mother start this beekeeping journey. I love honeybees, beekeeping and my AZ hives. We took classes, read all we could find, and I worked at a club apiary. At the apiary I soon discovered that the Langstroth, Top Bar and Warre' hives were not for me. Lifting and my "klutziness" were an issue. After I read an article about Brian Drebber in the "Mother Earth News" magazine on the AZ hives, I was hooked on AZ hives. I tried to find information, including ways to purchase, or get plans. I communicated with the few beekeepers that had AZ hives in the United States. All were on the east coast of the U.S. and information was very limited. Purchased hives were too costly to ship, so I then decided to design my own hives. I was lucky to have a daughter, Michelle, that was a structural engineer and my husband, Dean, also an engineer to guide my technical issues.

George Purkett, my bee mentor and Master Beekeeper, encouraged me and also gave me valuable advice with bees and beekeeping. I then shared information about AZ hives with others thus infecting Dana Schack, a friend, with the AZ hive bug. Dana soon was designing and building also. As we got the AZ hives off and running more people wanted information. Dana and I started giving presentations to beekeeping groups and then formed a Facebook group, AZ Hives-Pacific Northwest with the goal of sharing what we had learned with the Pacific Northwest beekeepers. Many joined our AZ hive family to learn, not just from the Pacific Northwest, but beekeepers joined from around the world. Soon I was writing articles to answer the many questions on how the AZ hive worked. The articles were written over a span of several years. These articles are now the basis for this book. It is intended for education with general information on AZ hives. I also include a lot of basic beekeeping information within the context of operating an AZ hive. To enhance understanding, I have included many of my photos and drawings. Finally, I wanted more information, so I took a trip to Slovenia to learn more about AZ hive beekeeping. Many of the beekeepers I saw there are in the reference section and the history chapter.

I do not claim to be an expert on bees or the AZ hive, just someone who had a hard time gathering information and wishes others to not have so much trouble. I hope you find use for the information I have shared here. It is only with my limited knowledge, perspective and skills. My designs are basic drawings as I'm not builder or technical designer but a teacher and artist. However, they would be a very good starting point if you decide to build an AZ hive. Hope you enjoy this book as much as I enjoy working my bees and the AZ hive.

Debra by Sherry Purkett 2022

Acknowledgments

Thanks to those who have supported me with editing, information or emotionally. Thank you Dean Boyer, Melissa Baltzell, Michelle Boyer, George Purkett-Master Beekeeper, Dana Schack and Paul Longwell-Master Beekeeper and the many friends that support me in my AZ beekeeping journey. Special thanks to those that took up reading and editing my book: Dean Boyer my constant support. George Purkett, Dana Schack, Paul Longwell, Dr. Dewey Caron-Master Beekeeper and Aleš Sűssinger all have read and gave me constructive comments as well as support. I could not have done it without their support and review of my book.

Bee by Michelle Boyer 2009

Dedication

I dedicate my book *The American AZ Hive* to Brian Drebber, the person who introduced me to the American AZ hive and invented the American AZ hive. He was the first to change the AZ frame size to fit Langstroth foundation in about 2015. He later coined the name "American AZ" for this version of AZ hives. He worked beekeeping with his granddaughter, Kylie, and brother Don to start and run Drebbieville Hives designing and building AZ hives until his passing in 2018. Brian was well known and liked in the beekeeping community. I especially appreciated his willingness to share information and his generosity. Thank you, Brian, you are missed.

Brian Drebber

Bee

Chapter 1
HISTORY

The American AZ Hive

Debra's beehouse

The AZ Hive uses a similar design to the Langstroth hive. It has frames, chambers (boxes) and the traditional bee space. The AZ hive enhances the design to make it easier for beekeepers to use frames with less disturbance for the bees.

Ellen at Debra's hives. 2024

Anton Žnideršič (AŽ) invented the AŽ hive in Slovenia just over 100 years ago around 1910. He based the AŽ hive on several hive styles, combined with the desires of the local beekeepers. This allowed for the use of a bee house as well as easy transportation.

AZ hives are kept in a bee house or similar building. The bee house protects the hive, bees as well as the beekeeper and equipment from the elements. Hives can be mobile if put on a trailer with a pod (container/beehouse). The bee house traditionally is also used to display paintings. The old paintings are a big tourist attraction in Slovenia. Each painting tells a story, or piece of history.

The American AZ hive is a variation of the Slovenian AZ beehive. The Slovenian AŽ hive has long been used in Europe. The Slovenian frames are larger. The American AZ hive's frames are the same size as the Langstroth frame. Thus, making them fit the same foundation and honey extractors commonly used in the U.S.

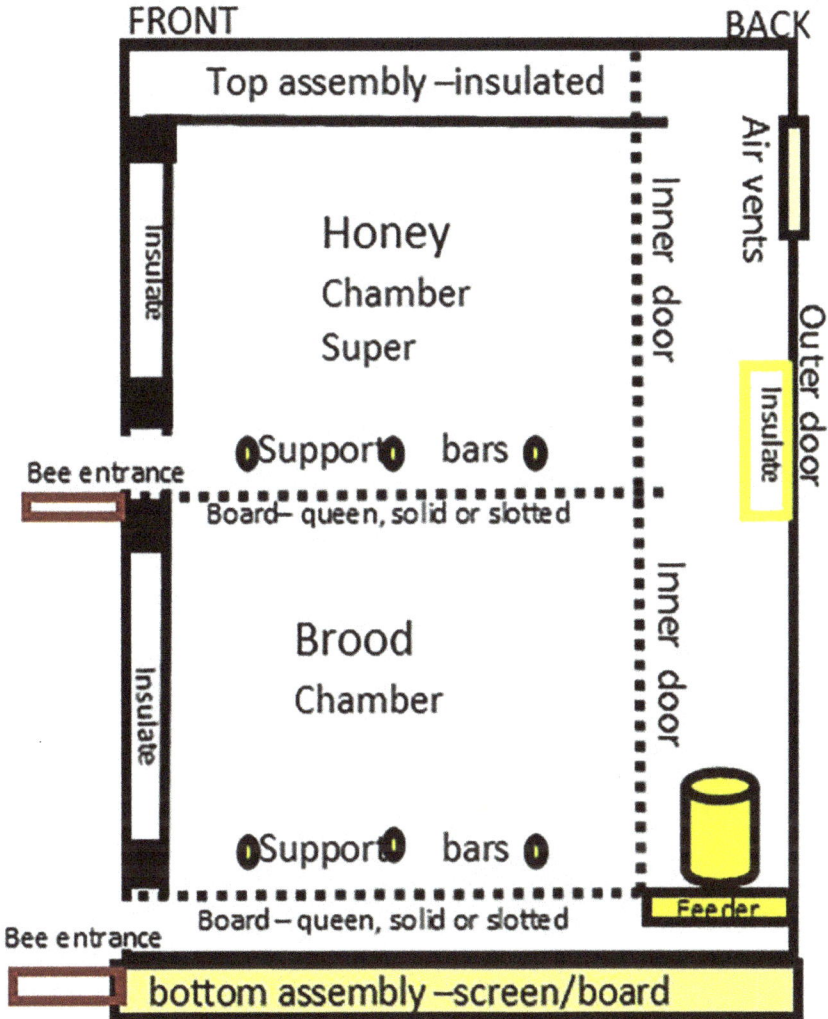

FRONT   BACK

Top assembly –insulated

Air vents

Honey

Chamber   Inner door

Super   Outer door

Insulate   Insulate

●Support● bars ●

Bee entrance

Board– queen, solid or slotted

Brood

Chamber   Inner door

Insulate

●Support● bars ●

Board – queen, solid or slotted

Bee entrance   Feeder

bottom assembly –screen/board

AZ hive drawing 2017

The AZ Hive is designed like a cabinet with the bees coming in the front and the beekeeper in the back inside the building. It consists of several chambers, each accessed with a separate inner door and a set of frames. Hives vary in the number of chambers from two to four with five to twelve or so frames each. The frames slide in and out on horizonal rods (bars) placed below the frames. The frames are much like books on a shelf. Between each chamber is a divider board to separate chambers. This board can be solid (to separate completely), slotted (for easy bee flow) or a queen excluder.

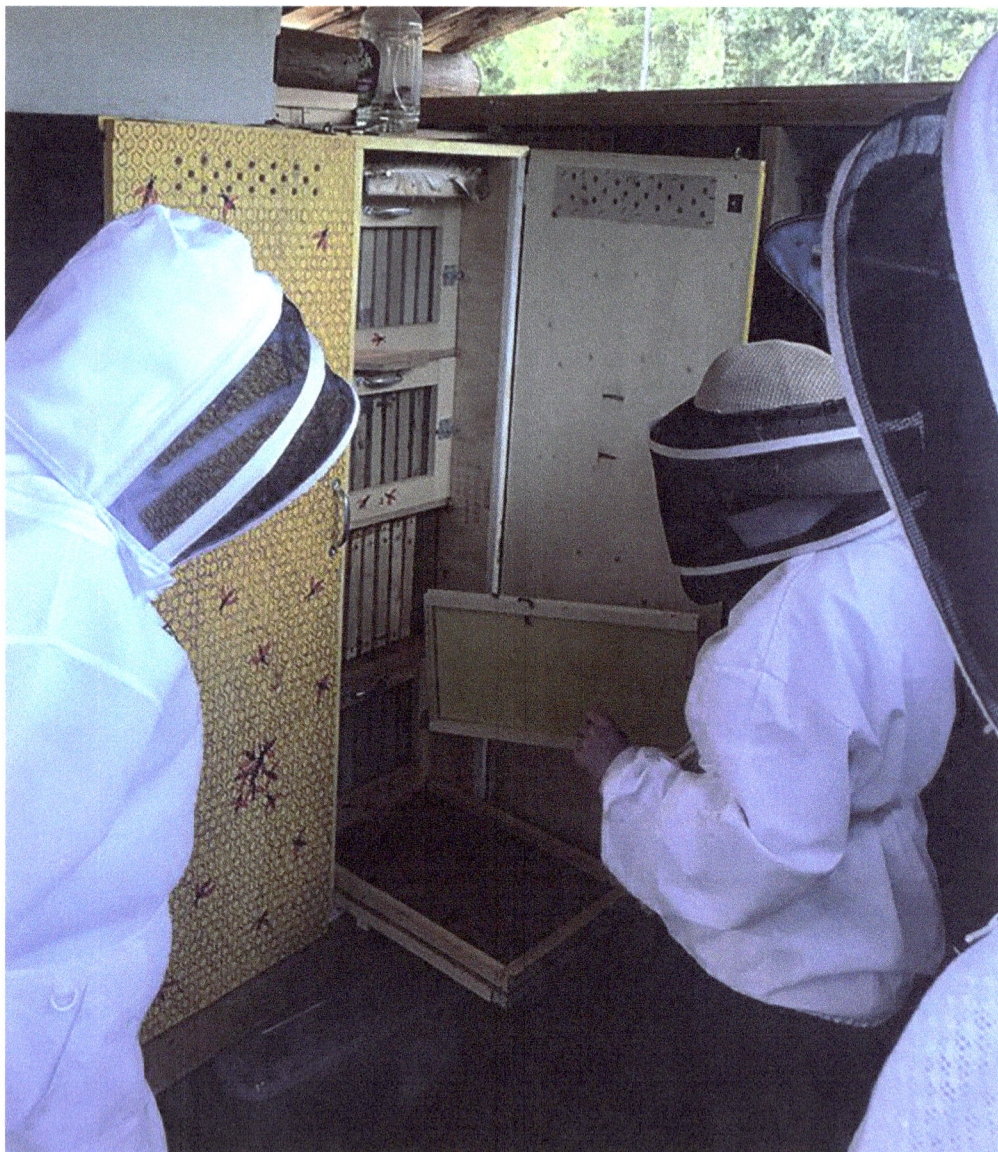

Inside beehouse Debra & guests. 2019 unknown photographer

The beekeeper accesses the bees from inside the bee house. They open the hive (cabinet) doors to the chamber they want and slide the frames out. The beehouse allows protection for the bees and beekeepers. The hives can be opened in the rain if needed. The beekeeper can choose which chamber to open without bothering the rest of the hive. Lifting is limited to one frame at a time. This allows elderly or handicapped beekeepers to easily work their bees (including wheelchair users).

Bee education. 2024

Bees are bees. Naturally, bees propolis and add bur comb wherever they can. The frames only have a few contact points on the rods (bars) or frame spacers. Frame spacers are at the front and on the inner door of the hive to keep 3/8-inch bee space between frames. This makes removing the frames easier, unless bees also add extra bur comb. Then it requires a little more touch. Sliding the frames to the side after removing the frame next to it usually disconnects frames easily. If you do not tend your hive and the bees build propolis or bur comb in the wrong place it is difficult just like a Langstroth hive.

The AZ hives are worked by moving frames, not adding new chambers to the top. The hive still has the brood chamber on the bottom and honey above. Feeders can be added to the back between the inner and outer door. Ventilation is done with openings at bee entrance and air vent on the outer back door. Moisture always needs to be addressed, the same as with a Langstroth hive. Landing boards are often added to the outside at bee entrance. For mobility the landing boards can close to seal the hive.

The American AZ hive has not been in the U.S. too long but is taking hold as more find the benefits of this style of hive.

AZ Hive History

A glance at the History

Beehouse with bee boxes. 2023, photo by author Beekeeping Museum

The AZ hive did not just appear. It grew out of hive history and is continuing to develop. This is a brief look at where the AZ hive came from and how it has continued to develop. Because its history is not always well documented, I've found inconsistencies and lack of information. This is a glance at what I've found.

Log "gum". Photo, courtesy Gene Kritsky

To start, we need to go back in time before the AZ hive to see why it developed the way it did. Then we can look at the way the AZ hive has changed through the years. It all starts in Slovenia but continues around the world.

Early on people just found hives in trees to get their honey. Hives were simply torn apart. But with time people began to care for bees and provide homes for them (hives). The 6th century beekeepers had bees in logs "Gums" that they could carry with Carniolan bees. They then began to place the gums in shelters for protection. Wooden bee houses came after that at farms.

Log "gum" 2023, photo by author Beekeeping Museum

House & Bee box paintings. Office at Slovenian Beekeepers Association

Around the world beekeeping was developing. Boxes (Kranjič) for hives become common. The first school for beekeeping was started in Habsburg court in Vienna around 1770. A painter

Stamp Anton Janša 1737-1773, PTT Jugoslavija, 0,80 Din. Paper, dim. 3 x 3,7 cm, Stored by the Museum of Apiculture, Radovljica Municipality Museums, Inv. n. ČM 967. Scan (n. ČM 117), 2025.

The photo the Portrait of Anton Žniderši, (black and white photo, unknown photographer), photo n. MF 1.

from a peasant family, Anton Janša, (Slovenian) was asked to teach. He is known as the father and pioneer of modern apiculture. For example, he is known for stacking the hives and painting the front panels. World Bee Day is set on his birthday, May 20, in his honor. Beekeeping was becoming a way of life in Slovenia.

Hives and the places Slovenians kept them continued to develop. In the 18th century painted panels in front of the hives were added. The paintings created stories and art on panels for all that passed by the hives. This painting on AZ hives continues today. The old painted panels have since become a tourist attraction.

Many hive designs were created around the world during the 19th century. Each trying for the perfect hive. Langstroth's hive design began to be seen around the world during this time. It was brought to Slovenia. Anton Znideršič (1874-1947), a businessman that loved the honeybee. He was very interested in beekeeping and hive design. He looked at many designs from the time. It is disputed as to what ideas came from which hive designs or beekeepers. Several hives and beekeepers - Kranjič, Langstroth, Gerstung, and especially Alberti's, are some believed he took ideas from. Anton wanted a hive that could be moved to fields, be protected from weather, house bees to produce more wax and honey in their harsh, cold, rain and snow climate. Each of the hives available did not meet his needs. He needed one that fit the needs and wants of beekeepers in Slovenia. My belief is he took a bit from many designs and his own ideas to make the AZ hive. It was more like traditional Slovenian hives. This would include being mobile, fitting in bee shed or house, and having the traditional designs (paintings). Having it horizontal with frames sliding out was key to stacking in beehouses. Around 1910 he had his new AZ hive.

Anton Znideršič beehives. 2023, photo by author Beekeeping Museum

The AZ hive is named after Anton Znideršič. I also see it named after Adolf Alberti (Germany) as well as Anton Znideršič. The name AZ hive can be different depending on where you are or who you are with. Names I have found are : AZ, Slovenian AZ, Traditional AZ, Alberti- Znideršič, American AZ (frame size same as Langstroth frame), rear loader (Holland), backdoor hive, L-AZ, AZ Langstroth (hybrid that takes Langstroth frames – I do not recommend this).

Over time beekeepers have created many changes for the AZ hive. Most AZ hives are in Beehouses and not mobile. Mobile ones are on trailers with beehouse pods (containers). I see various amounts of chambers. Europe tends to have 2-chambers and U.S. 3-chambers. The United States has had several people exploring the use of the AZ hive. Brian Drebber did a design change with the size of the frames in about 2015. He made the AZ frames the same size as the Langstroth frame creating the American AZ hive. Many of us followed his lead. I see various tweaks and did some of my own. I've changed the size to accommodate quart feeding jars. Dana Schack designed them, so each chamber was separate and then connected as needed. He also designed a five-frame AZ nuc. There are many more little details people have added or changed to fit their needs. Some will work, others not. I encourage you to think and try some new things. I've tried and tested several things with my hives (see in design chapter). This is how we got here with the AZ hive in the first place. Who knows what will come next.

AZ Hive Paintings

Bee panel painting. Tigeli 2023

Historically AZ hives have had various paintings on them. Exactly when, where, why and who started this is up for debate. The first one that I know about is in the late 1700's. The 18th and 19th century were active times for front board paintings. Several depictions are common – religious, life at the time, stories, lessons, and fantasy. Today the paintings vary widely from one color, geometric designs, detail objects, to entire mural scenes. What is known is beekeepers and the public enjoy the art. Visitors to Slovenia come to view the over 600 motifs from the past. Beekeepers today continue the tradition of painting the front of the AZ Beehive.

Why put paintings in front of beehives? Why it started we do not really know. But we can still see many of the early works. We know paintings identify specific hives for beekeepers and the bees. We know art sends messages to those that see it. We know people like doing artwork. Enough reasons for me. What do you wish to leave the future beekeeper with your hive? I put the mountains and plants that are around me on my main hives. Mount Rainer and the Olympics. Other hives get either a message (education) or how I feel at the time. Plants, trees, flowers or colors.

Art follows history and thus the life and feelings of the artist. History plays a huge part in what many artists will design. It influences their thoughts and feelings. The times or commissioner of the art may control what is allowed to be designed.

Bee panel paintings. 2023, photo by author Beekeeping Museum

The early paintings in Slovenia were done on front boards (Panjske Končnice). These were displayed on the front of the beehives. The country of Slovenia had, and still has, many beekeepers. Most farms would have a bee house with front panels. People passing by could view them. Many seem to have a message.

Čopi - panels, Debra, Debby 2023
photo by Barbara

A quick look at paintings in the world and Slovenia can reveal some of what was happening in that time.

▶ The 16th century (1500-1600) Slovenia had a Lutheran Protestant Reformation, and it was the time of the Renaissance in Europe.

▶ The 17th century (1600-1700) was the Baroque time with science and philosophy. The Academia of Science and Art was formed in 1693. American colonies started.

Hunter funeral, 1876. Animals taking the hunter to his funeral.
Many panels show animals and interactions with people.

Farming, 1876. Farmers plowing fields.
Day to day life in the 1800's required regular work. Panels often depicted this.

Mary with Jesus. Many panels are of religious nature.

Catching a Swarm" shows a group of men catching a swarm. Note the one smoking a pipe to calm bees. Swarm catching is only a bit different now. Date looks like 1843.

▶ The 18th century (1700 -1800) was the age of Enlightenment and Reason. Noted for peace. Except, Napoleon was active then and America was new. Anton Janša was teaching apiculture and educated as a painter.

▶ The 19th century is known for social change, industry (phone and electricity new), and the Victorian era in Britain.

Slovenia & the Honeybee
A Timeless love

Tigeli Bee on green flower

Slovenia, while a relatively new country but has always had a love of beekeeping. They keep only the Carniolan or "Gray" bee.

My beekeeping trip there in the spring of 2023 highlighted the love of bees and helped me study the use of the style of hive my AZ hives are based on. While not everyone is a beekeeper or knows about bees there are a large number of beekeepers. The capital has 125 apiaries & 800 hives. There are 214 bee groups in Slovenia. They have a rich history of beekeeping that continues today.

The country of Slovenia is made up of mountains, forest, farms and a little bit of seacoast. It is developed and modern but still full of nature, culture & history. The people are warm, welcoming, and generous. It is a place to go relax, experience outdoors, nature and bees. I found the beekeepers and the people there much like at home. I was lucky to experience some wonderful farmhouses, feasts, museums, caves, historical sites and things like the Lipizzaner horses, besides the apiaries. The country was amazing! Even with the heavy rains and flooding while I was there, we had a few sunsets, some good views and apiaries. The rich cultural information extended to the bees and their history in Slovenia.

Soca River

Nova Mesto floods

Dragon Bridge, Ljubljana photo by the author

Olimje

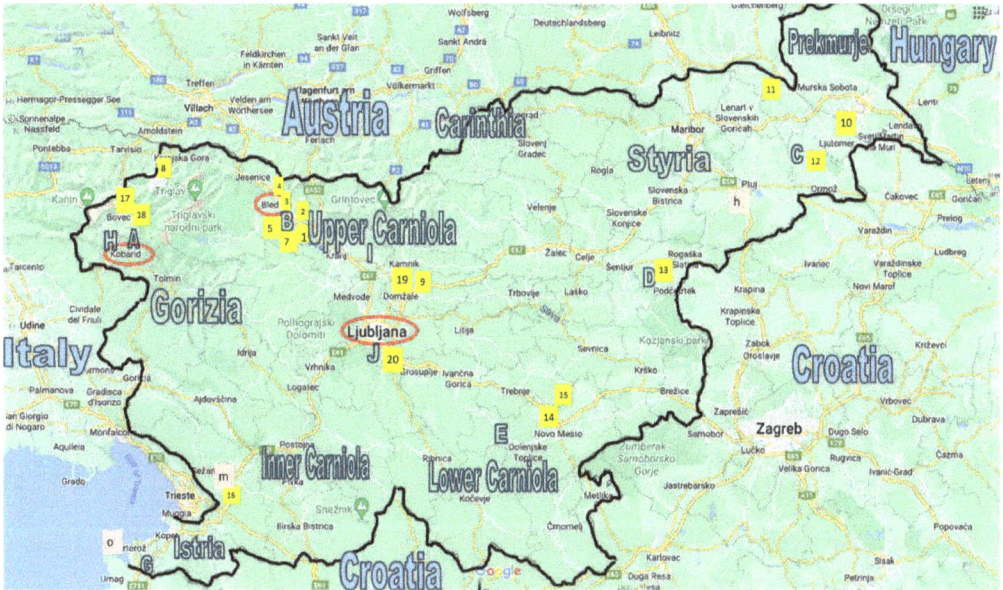
Map of Slovenia

Like all over Europe, there are many gorgeous buildings, castles and churches. There are historical sites from Roman and Napoleon times to WW I & II. Slovenia became their own country in 1991 after being ruled by many other nations such as Italy & Yugoslavia. They are a democratic republic with a parliament and prime minister. It is considered a social state. Different regions, make up the Slovenia country. (see map)

I found beekeeping history all over Slovenia.

Tigeli Bees

Skep Hive Čajnica

Snacks Bee Toni

Along with the Carniolan "Gray" bees (still the only bee there) items from the past can be found all over Slovenia.

6th century Middle Ages (500-600) had hives traveling carrying logs "Gums" & skeps. Logs were placed in shelters for rain and snow protection.

Log hives. 2023, photo by author Beekeeping Museum

Tigeli

Anton Jansa beehouse

Marjeta at Bee Toni

16 -17th century had wooden bee houses at farms. Boxes (Kranjič beehives)-stacking.

Old beehouses with various types of hives are abundant as are bees and gardens.   It seemed like every beekeeper had old hives from family or acquired somewhere. Tigeli is the round one (19th century) with Berlepsch style hives. Each apiary was a museum and modern apiary at the same time.

Old hives were at museums, apiaries, a coffee shop and the market (using grandpa's hive for storage). Or in garages, and old beehouses. Sometimes set up and sometimes just stacked places to store. Treasure troves everywhere! Hauzer House, Tigeli, Matija Komac, Čajnica, beehouses & apiaries were some I found.

Bled Market Lesnikinin Med

Hauzer

Hives stacked Matija Komac

Skep Hauzer

Bee Hive box. 2023, photo by author
Beekeeping Museum

Backpack hives. 2023, photo by author
Beekeeping Museum

Bee Toni - artist Alenka Peternel Hubert

Today there are new designs as well as the old ones for panels and murals.

Cart with beehives. 2023, photo by author Beekeeping Museum

These old hives were not just used in beehouses but traveled to the best nectar locations as hives do today. Early boxes on the beekeepers' back, and later, with carts.

Travel today is with a trailer (pod) so beekeepers can work bees in different places. 60 hives would not be uncommon per pod. The hive provides easy transportation with a truck and trailer.

Dollnšek Bee Pod - trailers

Simon Dollnšek Bee Pod (photo by Simon Dollnšok') and left Dollnšek Bee Pod inside

Bee trailer pods can be seen along the highways and roads.

Tigeli Bee Pod

Roadside Bee Pod airport

Skocjanske Jame beehouse at caves park

Tigeli bees

Beekeeping Education center of Gorenjska

Soca river areas

Honey extraction
Beekeeping Education
center of Gorenjska

Viewing seats Čajnica

Slovenian Beekeeping today has variety in hives and hive locations. AZ (common), also some Langstroth. Found in fields, parks, schools, roof tops, woods, homes.

Bee houses may include things like a honey room for extraction, a place to sit and enjoy bees, a bed for sleeping, viewing windows, apitherapy, and therapy equipment.

Health care using bees and plants is common in Slovenia (Apitherapy). National certification and education is required.

Beehive air,
Barbara Bee Toni

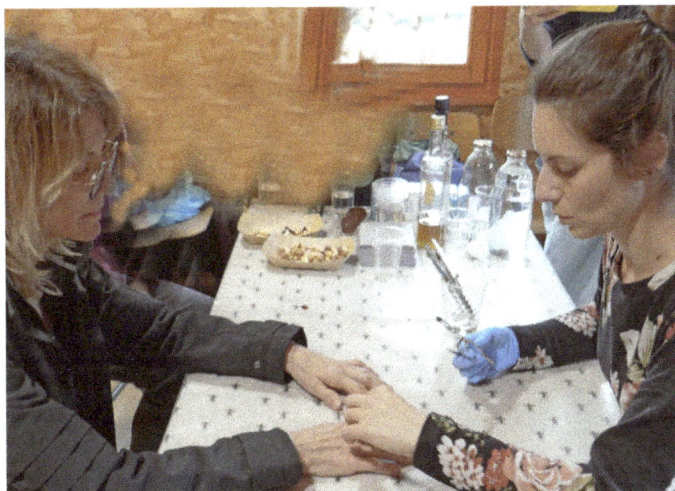

Apitherapy on Nedra at Gospodična Medična - Nika Pengal,

Gardens are also a major part with beekeepers. Many use herbs and grow vegetables for food.

Tigeli gardens

Bee hive bed - Gospodična Medična

Bee Tigeli

Bee Zavod

Roadside Beehive

Bee Toni nucs

The common honey is forest and floral.

Slovenian beekeepers have many rules to follow including national certification. Only Carnolian bees are allowed in the country, officials check apiaries, and the hive placement is regulated.

Modern equipment is used to keep track of everything from weight to temperature.

Matija Komac Beehouse

Bird panel detail. 2023, photo by author Beekeeping Museum

Bee Toni apiary

The Slovenian Beekeeper Association is very active.
Most beekeepers have a full-time job. Beekeeping is an extra.

Thank you for letting me share a bit of my trip to Slovenia and the honeybees.

Debra & Aleš

Plečnik beehouse in Ljubljana

Bee

Chapter 2
GETTING STARTED

Hive Choices

Which hive is best for you?

Honeybee's can use a great variety of places for their hive. Beekeepers provide hives for the bees as a convenience to the beekeeper. Different style hives are used successfully all over the world. The challenge is deciding which type of hive(s) you will use. Many places have laws that apply, for example in the United States we need to have some sort of removable frame (comb). Beekeepers also need to be able to examine the hive for disease. This article is just a quick look at some possibilities. I want to explore a few of the pros and cons of different hives, as well as questions the beekeeper needs to ask themselves when choosing a hive style.

A beekeeper that has been around should be able to help you with general bee care, while not all beekeepers will know exactly how every hive style works. They can help you check brood or check for disease by looking at comb and bees. They should know if a hive needs space for brood or honey. However, they may not know that an AZ hive difference from Langstroth is about moving frames not adding chambers (boxes). Details of how to work each hive is not universally known.

I cannot recommend that new beekeepers use any hive that has not been around for many years (50 to 100). Learn about bees first in a tried-and-true hive style. Then experiment. Some of the experiments do not work in the end. Bees will always behave like bees. The hive style is about how the beekeeper is able to access the bees and hive.

I only list four hives here, but there are many more good hives that I have not listed. They have frames or bars, vertical or horizonal, different amounts of space available for bees, mobile or not, made from various material or thickness, and other differences. But bees are bees and care

Debra by Vesna Sűssinger 2024

about their needs for food, safety, space and comfort. Most common hives meet these needs.

All four of these hive styles can work for a new beekeeper. Choose a hive style for your needs, not another beekeeper. Most hive types have potential adjustments to make it work for you. All styles have beekeepers that love that one type of hive. I recommend different hives for different people based on their different needs. Think about the items below before choosing a hive.

Bees wants and needs:

▸ Nice size cavity to build comb, lay brood, and store honey for winter.

▸ Protection from weather and predators.

▸ Food source.

George & Logzilla 2017

Beekeeper wants and needs: Some questions to ask and think about.

▸ Why honeybees? Products -Honey, pollination, wax or just love bees?

▸ Food – what will your bees eat? Look at local plants.

▸ Location -What protection do you need for weather (rain, snow, sun, wind) or predators (bear)? Where will hives be located?

▸ Finances – Beekeeping is expensive. Some hives initially cost more than others.

▸ Physical (your capabilities) – What shape are you in for lifting heavy hives? Or do you need an alternative method.

▸ Purchase equipment - Some equipment is easy to find others not. Build yourself?

▸ Assistance needed /available – mentor and classes?

▸ Aesthetics (looks) – What do you like?

▸ How to keep bees (time/treatment)– natural? Spend lots/little time working bees?

▸ Inspection – how often can you do it?

▸ Mobile – Do you plan on moving hives often?

▸ Space --number of hives, place to put hives

Debra photo and hives by Dana Schack 2024

Hives: alphabetical order -this is very general information

▸ **AZ (Slovenian or American) (frames and chambers)**

   • Has great protection (bee house). Cost more -last longer. No lifting -only one frame at a time. Few use AZ hive in U.S. Travel – trailer needed. Cannot add extra chambers.

▸ **Langstroth – (frames and boxes)**
Size of frames and boxes (deep, medium shallow – 5, 8 or 10 frame). Can add boxes to top for more space. No protection (can add shelter, lids, covers). Heavy lifting or moving one frame at a time per box. Easy to get parts. Common hive in U.S. Mobile with lifting and truck.

▸ **Long Langstroth – single box but long. Limited space. Lift one frame at a time. Not so mobile.**

▸ **Top Bar (bars one long chamber)**

- Limited space. No protection (can add shelter, lids, covers). No lifting – one bar at a time. Comb on bars, no support. Few use. Harder to move. Good natural wax when crushing for honey. Viewing window.

Some feel the top bar offers a Zen feel with beekeeping. Hive **must be kept level.**

▸ **Warre' (bars and boxes)**
  - Can expand for more space. Add new box to bottom. No protection (can add shelter, lids, covers). Lifting. Small boxes (about 8 bars). Comb on bars, no support. Can get tall. Few use. Mobile with lifting and truck. Natural growth for bees (top honey bottom brood). Viewing window. Good natural wax when crushing for honey.

Notes from Shaari, a friend, who has Warre' hives. Hive builds down. Hard for chemical mite treatment and inspection. Hive **must be kept level.** Correct comb as soon as you see it. **Quick look at these hives.   1 = poor 5=great**

Langstroth Bryan

Shaari's Warreé 2024

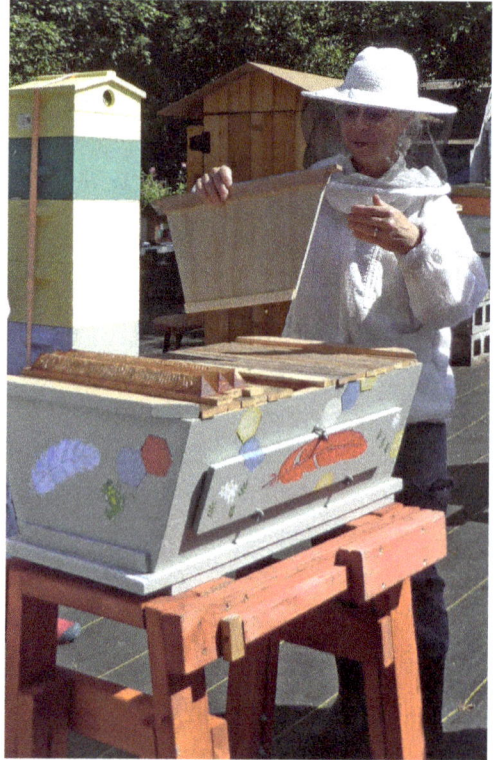

Top Bar, Shaari 2018

Hive	AZ	Langstroth	Top Bar	Warre'
1. Cavity size	4	5 (can add)	3	5 (can add)
2. Protection	5	3-4	3-4	3-4
3. Cost	3 -4	4	4	4
4. Lifting	5	3-4	5	3
5. Equipment	4	5	4	4
6. Mentor	3	5	3	3
7. Travel	0(in building) or 5(trailer)	4	2	3

Personal thoughts

I started with AZ hives some years ago. I spent a year learning about bees and beekeeping. I decided early on that the Langstroth hive was not for me. If I had to use the Langstroth hive I would not have bees. I was told over and over that the only hive you can learn about bees is the Langstroth –"You must start with a Langstroth hive". I saw that there were many other hives out there. They seemed to be doing fine.

Then I found the AZ hive -it was perfect for me. I especially liked that I did not have to lift heavy boxes and that the beehouse protected both hive and me. It was not an easy journey as there was little to no information available. Only one place to buy a hive (other side of country) and no plans. I decided to design and build my own. The good news is I had one longtime beekeeper say, "Go for it." He was supportive and interested in how the AZ hive worked. I got another new beekeeper also making an AZ hive (he says I infected him because I shared my AZ hive information). Many beekeepers were negative. Even had one refuse to sell me bees (he thought it was an experimental hive – I thought, not after 100 years or so).

I still hear beekeepers say you must start with a Langstroth hive. Or "I don't know how the AZ hive works so I cannot help you". You do not need to start with a Langstroth hive. The AZ hive has frames just like the Langstroth hive. What I learned at the club apiary, I repeated at home with my AZ hives. So, all this led me to write about hive choice. I believe we are all different and have different circumstances. Therefore, we do not all need to use the same hive style. Choose what is best for you from the start.

Buy or Build?
Slovenian or American AZ hive?

Decisions to be made

Like most questions the answer is "It Depends". Good, bad and faults are had by all. Listed below are some I've come up with. Things to think about as you make your decision.

Buying a hive costs more and may have less choices in design. Buying is also easy. As the builder, the hive is cheaper and lets you have any choices you can build, but building is a lot of work. The hardest things are finding hives, builders, plans, parts or exceeding your skills. More on where to find information, plans and ideas are in chapter on Plans or References.

	__Buy__	__Build__
	No work	Lot of work or a fun project
Cost Hive	Expensive	Low cost
Cost shipping	Expensive	Special parts only
Plans	Few options	Buy or design your own
Builders	Hard to find	You
Parts	Builder has	Find and buy (shipping)
quality	Good or bad (builder dependent)	Good or bad (builder dependent)
details	Fewer choices	Decide everything. Get what you want or can do (skill/tools)

Next you must decide what style of AZ hive to get.

Slovenian or American AZ hive?

The AZ hive, like most hives, was designed to have a specific style frame. The frame should be easy to bring in and out of hive. AZ frames have concave top and bottom bar to help propolis or bur comb not attach it to the round rod/rail. They sit in frame spacers at each end. The Slovenian and American AZ hive have different size frames. The Slovenian is taller but

(AZ) Slovenian, American,
Langstroth Frames 2020

2-chamber 2020

not as long (16 by 10 ¼ inches), and does not fit all honey extractors. The American is the same size as a Langstroth frame (18 by 9 ¼). Langstroth foundation fits the American AZ frame.

I cannot recommend using Langstroth frames for more than a week in AZ style hives. It tends to get stuck with propolis or bur comb (frames together, on rod, or in front of hive without a way to disconnect them. The issue is the wings, width at top side of frame (touching each other) and lack of concave bottom or top. (more on frames in chapter 3 "AZ Frames") Langstroth or Hybred AZ hives are for sale. They have not done well for those I know, due to the bur comb and propolis build up causing the frames to not be able to be easily removed.
Decisions to be made are not over yet. How many frames per chamber? (8-12?). Number of chambers? What type of feeders? Inner door screens? The list of decisions goes on whether buying, building or even after which frame style. Learn more before deciding what you wish to do.

There is currently not a standard size for hive or frames. This causes issues with purchasing supplies.

**Make sure new designs have been tested by bees for several years.

Adding bees to AZ hive

Swarm moving in AZ 2024 by Dana Schack

Swarm in AZ 2024

Swarm, Package or Nuc?

Placing new bees in an AZ hive can be easy. If possible, place the queen in the hive – bees will follow. You can dump bees on a hive table to crawl in, in an empty chamber above frame with queen, or use a chute to pour them into the hive. (more information Chapter 5 "Queens in AZ Hive")

Swarm

Depending on how you captured the swarm it could be dumping or if you used old frames putting frames in the hive. The easiest way is for the bees to just move in. Swarms need some empty frames to build on, as they are ready to form a new hive. Photo is of swarm moving in with no frames, it was a mess, but not my hive so I had fun.

Package

With a package – move the queen (keep in cage, trade cork for marshmallow) to the hive. Connect to a frame, remove one frame to allow space for queen cage. Now, dump bees into the hive. Leave the package box in the chamber above so all the bees can come out -remove in a few hours or the next day. Check to see if the queen has left her cage in a day or so. If not release (after checking to see they have accepted her -- feeding queen =yes, attacking queen = no leave longer).

Plastic Package 2022

Wood Package 2022

Nuc

The nuc would be easy if it is on AZ frames (not likely). Just put into hive from nuc. If on Langstroth frames you can cut off Langstroth frame and put foundation into AZ frame. Or use a Langstroth hive and slowly move bees and brood to AZ frames (long time -weeks). For details on frame conversion in chapter 2 "Frames Langstroth to AZ".

Cardboard Nuc 2017

Tools, frame, saw 2019

Frames Langstroth to AZ

Changing from Langstroth frames to AZ frames

Starting an AZ hive with Langstroth frames is more of a challenge than starting with AZ frames. But it is not only possible but not too hard. After designing and building my AZ hive I was ready for bees. I decide that the best way to get bees was a five frame nuc, I found two local locations to get 5-frame nucs. I carefully researched all of the people who had moved from Langstroth frames to AZ frames. I found one person in the country, Brian Drebber. He popped out the foundation and then popped them in AZ frames. That did not work for me. Next with assistance from my mentor, George Purkett, we tore the Langstroth frames from the foundation. That was very messy and difficult. That did not work well either. I then tried the cut off method – that I made up. I've settled on the cut off the Langstroth frame and put the foundation in an AZ frame method. It works best if you have wood Langstroth frames with plastic foundation. Other types of Langstroth frames and wax foundation work but are hard or messy. These are the steps I take:

Brush off bees 2019

1. Have AZ frames ready. I build my own frames. I only connect three of the sides. The bottom and two sides. I leave the top open. This allows me to slide the foundation into the frame. Then I add the top bar.

2. Remove bees from the Langstroth frame. This makes it so much easier. Brush bees back into the nuc box.

3. Cut around the edges of the Langstroth frame to loosen any wax or propolis from the wooden frame. I just use a knife or hive tool.

Cut edges 2019

Cut top of frame 2019

4. I use a saw (any small power saw will work). I cut the top bar of the Langstroth frame. Either in the middle or on each side. Now the top bar is not connected.

5. Pull the top bar off the frame and foundation.

Foundation out 2019

Insert to AZ 2019

In AZ 2019

6. You can now pull the sides off the foundation. Or just pull down.

7. The foundation can now be pulled/lifted off the bottom bar of the Langstroth frame.

8. Slide (more or less) the foundation into your waiting AZ frame.

9. Next add the top bar of the AZ frame and nail down. Glue if you want. You now have an AZ frame ready to put into your hive. Put frames back into nuc box to collect the bees.

10. Once you have all the frames done you can add them to your hive. The process takes a while but you get faster with practice.

Add top bar 2019

Place in hive 2019

Plastic frame 2019

Wax wired foundation 2019

Problem frames:

Plastic Langstroth frames. I cut off the edges leaving a foundation much the size of regular ones. I then added holes in the plastic foundation. I used wire to connect it to the AZ frame.

It is very messy.

Wax foundation in the Langstroth frame. I cut off the frame (carefully). Removed the wax (with wires). It got rubber banded into the AZ frame. Fragile.

Special thanks to Dana Schack and Sean. I photographed while they changed Sean's frames. Most of these photos are of them working.

Bee

Chapter 3
THE HIVE

AZ Hive

Debra hives by Sherry Purkett 2022

The AZ hive is one of the elements that make the whole AZ concept work well. The other elements are the frame style and the bee house (or shelter for the hive). Each one can be designed and constructed a little different depending on what you want or need. The hive contains all the key parts of the AZ system much like a body holds all our parts.

The AZ hive, like many hives, starts with a box like structure. It has a top, bottom and sides. The front has holes for the bees to come and go. The back has a door for the beekeeper to access the hive. It looks much like a cabinet you might hang on the wall in your kitchen.

The details of the hive give it form and function. There are decisions to be made when buying or building. How big do you need the hive? How many chambers and how many frames? The way the frames are held secure in the hive with spacers and rods is important and can have some choices. How do you wish to feed your bees? The base concept is the same as for any hive. All decisions need to keep the bees, beekeeper and environment in mind. For instance, the 3/8 inch bees space is needed throughout the hive as bees will fill what is left.

George with Nucs by Sherry Purkett 2022

2-chamber hive 2020

1. The size of the hive.

Chambers and frames control the size of the hive. If you wish more space for the bees add chambers and go with a ten-frame chamber. If you wish something smaller, like a nuc, go with one chamber and five frames. It is a personal preference. The brood chamber(s) are usually on the bottom while honey is on the top.

a. *Chambers*. Many hives in Slovenia and Europe have two chambers while in the United States many have three chambers while a few with four. I made this decision based on number of boxes Langstroth users have near me. Others decide what will fit into their bee house or how many hives they want in the end. The smaller hives require a little more movement of frames. The larger ones can be divided into more mini hives with divider boards.

b *Frames.* The number of frames per chamber can be anywhere from five (nuc) to ten or so. Some prefer to run the ten frame and others go with an eight frame.

3-chamber hive   2020

4-chamber hive   2020

2. Holding frames.

What holds the frames in an AZ hive are the bars or rods under the frames and the spacers on the front and back of the hive. Both rods and frames spacers with the frames are designed to have few connections. Bees do like to propolis everything.

a. *Bars/rods.* The rods under the frames hold their weight. Generally, there are three evenly spaced, 3/8-inch rods. The rods are round to provide less connection to the frames and the frames have a concave top/bottom or the same reason. The rods are connected to the side wall of the hive. They can be removable.

b. *Frame spacers.* To maintain the 3/8 inch bee space there are spacers at either end of the hive. Two each are connected to the front wall and the inner door. They hold the top and bottom of the frame in place. There have been a variety of things used such as nails and pegs but most use a metal piece. This metal is cut with triangle points between the frames and a small frame rest at the back. The design is meant to have less connection points for the bees to propolis. It has a L shape to add screws to the wall or inner door. These dividers can be purchased commercially.

3. Divider boards.

To separate the hive chambers, divider boards are used. These must be in the hive with bees. Bees will fill the space if no divider board is in place. They slide in and out on top of very small metal (or wood) L

Inside-rods, spacers, frames 2020

Right: Drawing side view 2016

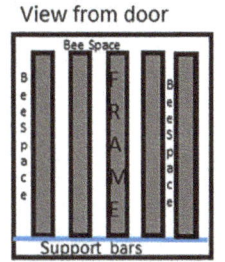

View from door

Drawing Frame spacer 2016

Board –queen, solid or slotte

Inside & divider board 2020

Inside & queen excluder 2020

shaped rails connected to the sides. They sit just below the rods. There are several variations of these dividers (solid, slotted, queen excluder). The bees love to propolis and build bur comb between chambers. They start at a frame and move through the divider board in the spring with drone comb. I've started adding drone frames near this space. It works great.

a. The solid divider board is used to completely close the space and not allow bees to move between chambers. Add an L shape metal to the front of the hive (like on sides). It helps close the end space.

b. A slotted divider board is used to allow the bees to flow freely between the chambers. It has many 3/8 holes running mostly the length and width of the divider board. The size of hole, length and width can vary. I have been experimenting with round holes instead of the long slots – so far it is ok.

c. The queen excluder divider board is used to keep the queen on one side or the other. Often to keep the queen from laying in the honey chambers. I have had two queens in separate chambers of one hive with a queen excluder between them.

Landing board 2019

4. Entrance bees/beekeepers.

Bees and beekeeper need access to the hive. The bees have their entrance on the front of the hive and the beekeeper on the back.

a. *Bee entrance*. The bees have entrance holes at the front of the hive. Each chamber can have an entrance with the main one being at the bottom of the hive. Most put the chamber entrances at the bottom of the chamber where the rod sits. The size of the entrance can vary. I used 10" by 1" for the bottom and 6"by 3/8" for each chamber. Each entrance should have entrance reducer and way to completely close the entrance. Being able to close the hive is especially important if you travel with the hive. I have plugs to put in the chamber entrances and a mouse excluder for the bottom with slides to close it up. I see some with flip up landing boards to close the hives. The round entrances discs (plastic or metal) with single hole, queen excluder or air vents work wonderful for upper entrances.

b. *Beekeeper entrance*. The beekeeper generally has two doors to enter the hive.

i. The first door (outer door) is solid and it opens on hinges like a kitchen cabinet door. It allows the beekeeper to have a space between the bees hive and the door to put feeders, insulation and check bees without opening the hive. It is nice to have this door on hinges

Landing board Gospodična Medična 2023

65

Inner door Plečnik Castle hive 2023

Inner & outer doors 2020

that allow you to lift off the door and set aside. Include air vents that can be opened for more ventilation. Put screen in the air space as well as a cover to close off excess air flow.

ii. The second door (inner door/window) allows access to the inner hive. This inner door often has a window (plexiglass or screen) to view the hive from outside or allow air flow. Some even have the window removable with the inner door a picture like frame. This door normally slides in and out with some type of clips, pegs, or bars keeping it locked shut. The bees side of this door has affixed frames spacers at top and bottom to keep the 3/8 bees space with the frames.

3. Bottom of hive.

The bottom of the hive varies as to what people do. It can just be a solid bottom. Screen bottoms are used by many now. I have an extra space for bees of ¾ inch. This is a hangout on hot days, place for Oxalic Acid vaporizer, place for pollen patty and place to use for dead winter bee clean out. I have never had bees build here. Mine has the screen under it with a drawer/board to close off screen and collect anything the falls in the hive (dead mites, pollen, wax capping …). This way I have another look at what hive is doing.

AZ Frames

There are several elements of an AZ hive that make it function well. Such as the bee house (shelter for hive), cabinet style hive and the AZ frames. Variety can be obtained in many of these elements. I want to focus on the frames right now because they are the heart of the bees hive.

Note: There are many ways to do frames and this is only a look at what I have tried. There are also many different terms used – but I am covering what I'm doing. The variety of ways and terms are not wrong, just different.

Basic AZ frame:

The basic AZ frame has a concave top and bottom wood bar. This concave design allows the frame to touch the least amount of places while setting on the bar (rod) that the frames sit on. Since bees love to build (comb) and propolis (glue) less touching is better.

AZ Frame ends 2017

▶ Slovenian AZ frames and some American AZ frames have the side pieces with the concave cut out also as they are connected on the outside of the top/bottom wood bar of the frame.

▶ If making the frames, cut the concave part first, then cut into frame length. Add sides. Connect with staple, nail or dovetail joint. Glue or not glue.

▶ I recommend the concave top and bottom bar be the length of the frame. A slot can be added to the inside of the concave bar for plastic foundation. The ends (sides) are just flat pieces of wood that can fit inside the top and bottom concave sides. This allows you to nail from the top and bottom giving additional strength when pulling the frames out of the hive. This is a design by my daughter, Michelle Boyer, a structural engineer. She had decided the frames were not strong enough.

▸ All frame pieces are an inch wide to fit the spacers. The depth of the pieces vary depending on person making them but to get the concave part you need about ¾ of an inch. The ends can vary more. They could be ¾ inch or something like ½ inch. Whatever you chose, make sure that the frames fit your chamber. I find there is a little extra space to play with as needed if you get some different size frames.

Frame styles:

The frames in Slovenia AZ hives are sized a little larger than the American AZ frames. The Slovenian frames are 16 inches by 10 ¼ inches while the American AZ frame is 18 inches by 9 ½ inches. The Langstroth is 18 inches (plus 1 inch for wings) by 9 ¼ inches and the sides are 1 inch at the bottom and 2 inches at the top. As you can see the sizes make a difference to the chamber size and foundation fitting into the frame.

Frame size 2020

There are many ways to set up the frames. A variety of other things can be done, I just haven't tried them all yet.

▸ Bees can make all their own comb with just an empty frame. I find comb a little precarious using this method in warm weather.

▸ Bees wax foundation can be used. This is a traditional method. It is good to wire or add support in the frame for strength to the comb. Wiring is often done top to bottom through holes made in the top and bottom of the frame. The thin wire is strung up and down through the holes. Wax foundation is laid on the wire, heat added to melt wax on wire. Careful to not have too much heat as it will melt your wax off the wire!

Frame end style 2020

Bars Frame 2020

Drone Frame 2018

- Plastic foundation can be inserted into the frame to start the bees and give additional strength to the comb. Add a slot to the inside of the frame for the foundation to sit in. I often use black for brood and yellow for the rest. I add several holes in my plastic foundation so bees can flow between the frames easier. This is especially good in the winter. Currently I'm doing 5 holes – 3 near top and 2 just below and between the top holes.
- Consider a wood bar in the center, length way. This works nice for support.
- Part plastic foundation. Top 1/3 to 1/2 for foundation with bar added on the bottom side to hold foundation. Drone comb is commonly built on the bottom empty half. Nice way to allow for plenty of Drone cells and they fill with honey for the winter.
- Bamboo skewers can be added (2 -top to bottom or another way). Bees fill it fine but the bamboo leaves a bump at top and bottom of frame.
- Top bar triangle at top. Bees build up fine but like empty frame the comb is precarious.

Top bar style frame 2020

Langstroth to AZ Changing frame 2024

Changing foundation from Langstroth to AZ frame:

Nucs often come with Langstroth frames. This is not a problem, there are several steps to take. See chapter 2 "Frames -Langstroth to AZ"

The number of frames per chamber in hive depends on what you want and need:

AZ hives can have a variety of frames per chamber and a variety of chambers per hive. Traditional Slovenian have 2-chambers while I see most American AZ hive having 3-chambers. Most seem to have 10 frames per chamber, but I also see 8 frames. Then there is the AZ nuc with 5 frames. I have two 4-chambers, two 2-chambers. Both are 10 frames hives and two 5 frame nucs. I based my decision on location (how many boxes do local Langstroth users have), number of hives I wanted in my bee shelter and uses I have for them. The number of chambers and frames per chamber is up to you. What do you want? All are ok.

Inside rods, bars   2017

Rods or rails:

(mentioned in Chapter 3 "AZ Hives")

Each chamber has 3 rods to set the frames on. They are generally a round pole like bar/rod that can be metal or some sort of plastic as long as they are strong enough to hold several frames heavy with honey. They are round to avoid connection points for the frames (concave side of frame sits on these). They are evenly spaced under the frame and connected to the sides of the hive or removable with slots in the wall sides. Frames slide in and out on these rods like a book on a bookshelf. Rods are 3/8 inch to have bee space.

Frame spacers:

(mentioned in Chapter 3 "AZ Hives")

Frame spacers are used to keep the frames with 3/8 bee space. They are placed at the front and back of the hive. One set separated the top of the frame while another does the bottom. The front sets are screwed into the front inside hive wall while the back set are on the inner door. My frame spacers are metal. They are designed (triangle point between frames) to touch as little of the frame as necessary to maintain spacing.

Drawing Frame spacer 2016

Frames Spacer 2020

Frame holder/stand:

(mentioned in Chapter 3 "AZ Hive Accessories" and Chapter 6 has plans)

A frame holder/stand can be used when removing frames and you need a place to set the frame. It is just a couple of boards with frame spacers to hold frames.

Frames Spacer 2020

AZ Hive Accessories

There are several items that make it easier to work in an AZ hive. I have listed a few below. There are other things and ways different beekeepers work their hives.

Hive Table: This connects to the back of the hive under the frames. It keeps things, like a queen, from falling on the floor when you pull out frames. The tables can slide in at the bottom under the divider board or where the divider board goes. I'm thinking of making a new one that has legs for better support.

Hive Table 2023

Frame Holder/Stand: This is a free-standing place to put frames when not holding them. It is made with a couple of boards with metal frame spacers to hold the frames steady. I have two for 10 frames each. I use it when I need to have frames with comb, bees or brood in a safe place that is not the hive.

▸ I also have one that holds one frame.
▸ Langstroth frames have tabs that hang frames.
▸ (mentioned in Chapter 3 & 6 "AZ Frames" design plans)

Frame spacer: This is a board with the metal frame spacers attached to give frames in the hive the 3/8-inch bee space. Used to space the frames after putting them in the hive before closing the inner door.
(mentioned in Chapter 3 & 6 "AZ Hives", "AZ Frames", plans)

Long hive tools: Reaching the front of the hive from the back of the hive to remove things like dead bees requires a long tool. Several types would be good. One with a flat side to scrape bees and one flat and sharp to cut propolis and bur comb from top of frames.

Frame Stand Plečnik 2023

Frame spacer handle Plečnik 2023

Long hive tool 2020

Frames Spacer 2020

Feeder jars 2017

Divider board handle 2017

Smoke stick 2020

Rounded hive tool: A rounded end to scrape the concave parts of the frames removing propolis or bur comb is nice. Thank you, Dana Schack, for designing and making.

Feeders: In hive feeders are easier to monitor the food each hive is getting. I like three-quart jars with screens to keep bees in hive. Others style feeders are available. Some attach to inner door.

Divider board handles: cut handles into the end of divider boards to get a better grip when you need to pull them out. Bees will propolis all edges by the sides of the hive.

Smoke sticks: Smoke is often used in beekeeping. While I generally do not use it, there are times I love them. The smoke sticks put out a constant small amount of smoke. I have put them at the edge of the hive to move them back in. Thank you, Paul Longwell, for showing them to me. Smokers tend to use too much smoke in the beehouse.

Robber screens 2023

Robber screens: There are times when bees or yellow jackets wish to rob a hive. Covering the entrance with a robber screen helps the smaller hive bees protect better.

Quilt box: Box that fits inside chamber with moister absorption item (wood chips). Screen on bottom. I have quilt box set on top of 4x4 blocks in top chamber so emergency sugar can fit under it. Remove in the early spring or bees will use the space to build.

Quilt Box 2020

Foam 2021

Open Cell foam: Cut to fit between inner and outer doors as insulation and a moisture catch.

Nucs: Small hives (5 frame) are good for grafting new queens, starting a new hive or place for small hive of bees. Free standing makes it easy to move around for mating or moving small hive of bees. Designed by Dana Schack.

Nucs 2020

Magnify

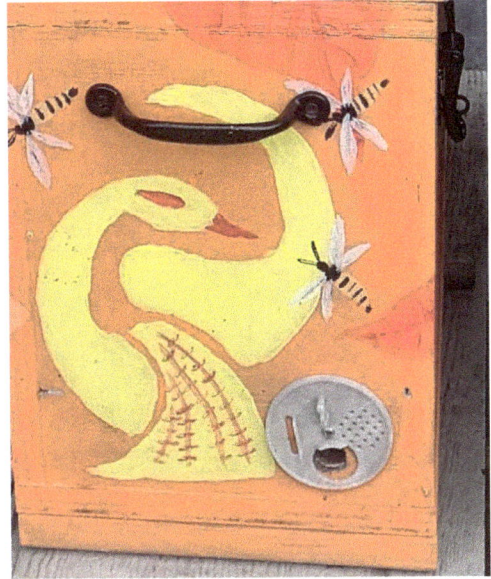

Handle, disk open 2020

Handles on movable hives (Nucs): I added handles on hives I need to move around. Helps in carrying.

Magnifying light set up: Helps in a dark room to see the eggs on the frame.

Frames: Several types of AZ frames. I keep ones with and without plastic foundation, drone (half no plastic foundation) and several experiments.

Divider boards: I have solid, slotted and queen excluder divider boards. I'm trying round holes instead of the long slots as they are easier to build. You can also have some with partial opening (end to make only exit passage, middle for winter feeding). Keep them in the hive anytime you have bees.

Pollen Traps: Pollen trap collection at bee entrance.

Chair: I have a couple of stools (tall and short) to sit while I work the hive.

Divider Boards 2020

Pollen Trap Tigeli 2023

Bee

Chapter 4
THE BEE HOUSE

The Bee House

The bee house is the last element that makes the AZ hives work well. With the cabinet style hive and AZ style frames it completes the circle. Like the other two elements of the AZ concept, it can have a great variety of styles. I have a simple shelter, while many in Slovenia have full bee houses. Here in the United States, there are those that have full bee houses to those that just have a mini shelter. Slovenia, Austria, Switzerland, Northern Italy and Germany all have bee houses.

Debra & bee house unknown beekeeper 2020

Bee Hut George, Ellen by Sherry Purkett 2024

The bee house is an essential part of the AZ system because it protects the hive from the weather. This is protection for the bees, hives, and beekeepers. The bees get extra insulation from the building in the winter. The bee house helps protect the hive from rain and snow. The hive itself is not designed to be out in the elements. The extra protection allows them to last for many years, even to the extent that they can be passed on to the next generation of beekeepers. The beekeeper can work in a protected area with all the tools stored right at the work site. I find it especially nice on a hot or rainy day. I can also check in the winter when needed or I wish to see my bees.

Anton Janša 2023

Bled Farm

Čajnica 2023

Styles of Bee Houses: There are many styles and sizes of bee houses for the AZ hives. The fanciest ones have sleeping quarters and a honey room. All accommodate the hives and most of the basic equipment needed to care for the bees. At the lower end of the bee house scale is the simple shelter that only covers the hives with a roof. Decisions need to be made as to the style and cost each beekeeper wishes. The size of the bee house needs to accommodate your needs. How many hives, honey room, storage for all equipment, running water, and the list goes on. I recommend building a little larger than you think you need. I have a shelter that is about 10 feet by 12 feet. It has room for a row of 5 hives across with space for four chambers high or two stacked two chamber hives. I have four hives and equipment. There is no excess space in my shelter once I

Olimije 2023

Slovenian Beekeepers Association 2023

have everything inside. The key for the AZ hive is you have a roof. After that it is a personal decision. The look of your bee house needs to fit you.

Roof: The front of the roof of the traditional bee house has an arch or overhang at the front over the bee entrances. This helps the air flow with the hives as well as protecting the front of the hives from sun, rain or snow. The distance it hangs out is directly related to protection from the sun in the summer on the lowest hives.

Bee Hut 2017

Bee escape (windows): If you have a closed in bee house you need a bee escape. While working the bees, some naturally come into your workspace. The bee escape should be a window up high to attract your bees by the light outside. Removing excess light sources help the bees to go for the escape. The top, front of the bee house is a good place to put the bee escape. It can be a one-way escape. If you have a shelter like me then they come and go from all over.

What should be in the bee house? Whatever you want and can fit. Storage of equipment, worktables, chair for working lower hives, something to stand on to work upper hives, hanging space, tubs to keep pests out of equipment, and lots of shelving are some of the normal things in the bee house. But you could have a luxury bee house with things like sleeping, apitherapy, honey room, freezer, sink or viewing space. At a minimum I recommend having some bee equipment to work the hives.

Inside Debra's beehouse 2020

Outside Debra's bee house 2017

Paintings: While not essential, traditional AZ bee houses were decorated with painted panels on the front of the hives. The designs vary greatly from stories to lessons to learn. Today many of us just put a design we like. I enjoy making murals, so mine is of local mountains and flowers. These painted fronts also help the bees find their hives. Which reduces drifting.

Issues with the AZ hive beehouse: While there are many good things about having a beehouse there are some drawbacks to address depending on where you live. Hot, cold or rainy climates create changes inside the building unless it is a sealed climate-controlled building. I have converted an old sheep shed. In the hot areas the building may overheat – add a fan to the roof. Cold areas keep the building cold in the shoulder seasons. This can cause hives to not warm up as soon during the spring buildup. I added solar heat and fan to help. Or heat can be added (not above 40 degrees). Moisture in the air can make the inside damp, depending on climate. My area gets more than 100

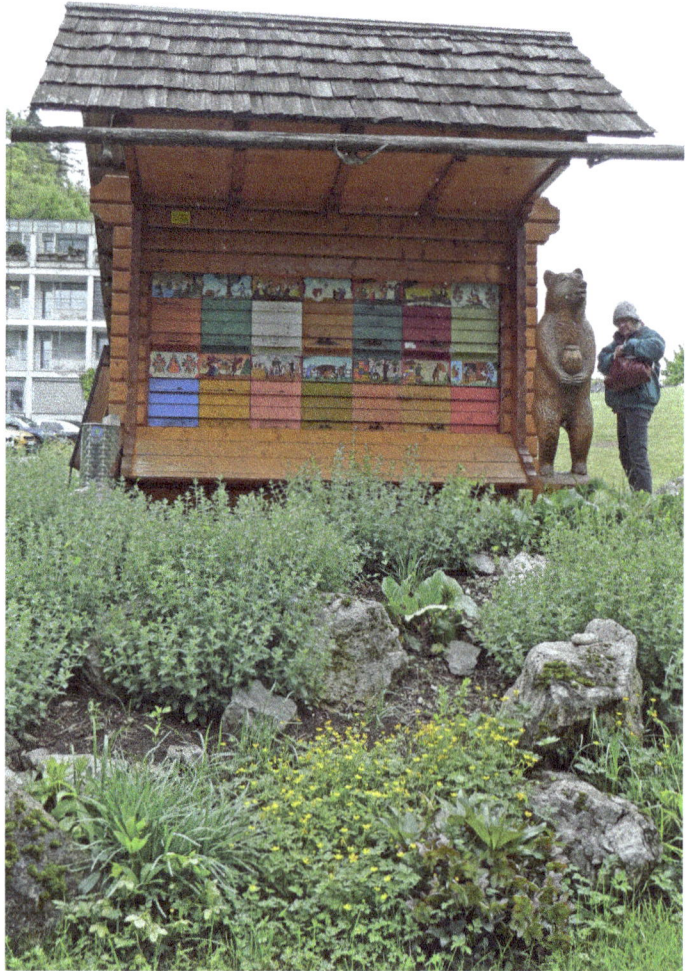

Bled park 2023

inches of rain and the air is wet even inside shelter. I've added solar heat, rubber floor mat, patched most air holes and insulation to help with each of these.

The bee house, like many of the elements of AZ beehive system is full of personal decisions from the beekeeper. There is no right or wrong answer but make sure you think of your potential needs before you start.

Bee House Traditional Drawing

General design for beehouses is a slanted roof on the front for sun (varies with location). Bee hives are located to get plenty of sun in winter and less in summer.

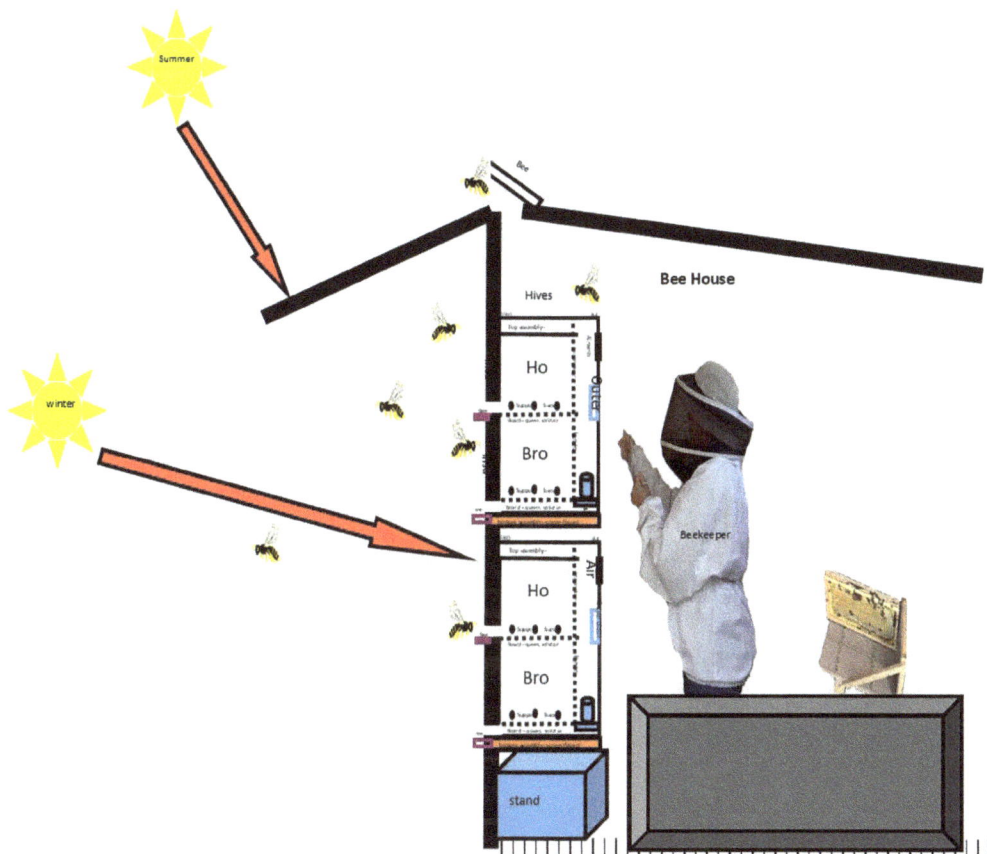

Bee house Drawing 2020

Debra's Bee House Shelter Drawing

Debra has a sheep shed that turned into beehouse.

Dimensions: Beehouse 8' x 10' (need 12x10). Hives 17"x24". Above 2-chamber hives/storage is open work/storage space. Hooks for storage all around. Short windows high on back and sides. Bee escape & shades for all windows. Solar power and heat (on roof). Roof overhang (need 3'). Stall mat on ground (need floor). Old log sheep shed conversion.

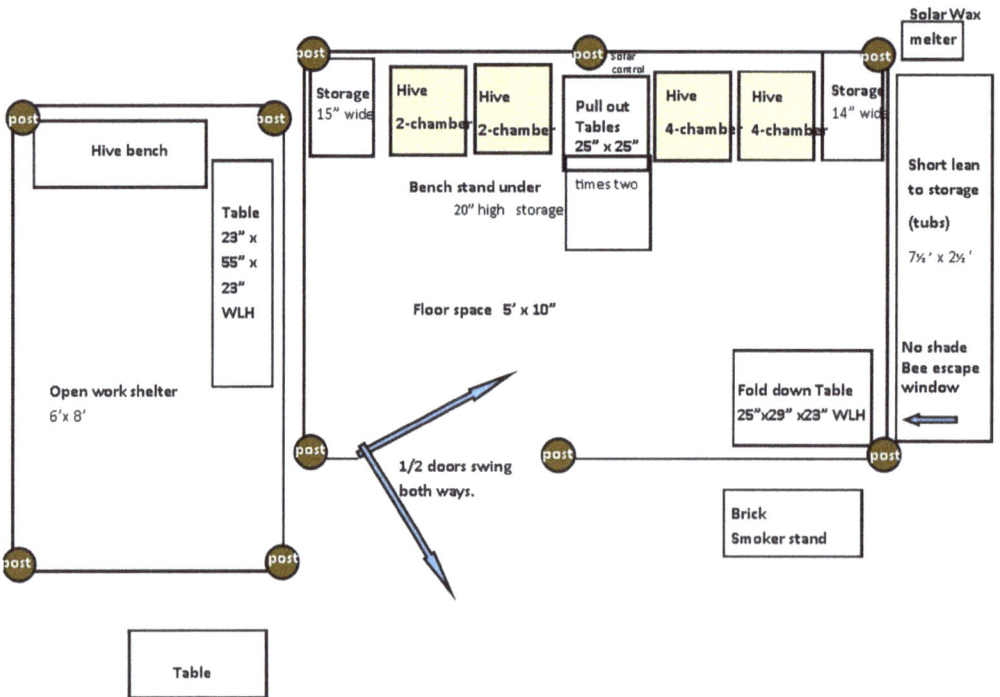

	Solar Wax melter

Hive bench

Open work shelter
6' x 8'

Table
23" x
55" x
23"
WLH

Storage
15" wide

Hive
2-chamber

Hive
2-chamber

Solar control

Pull out
Tables
25" x 25"

Bench stand under
20" high storage

times two

Hive
4-chamber

Hive
4-chamber

Storage
14" wide

Short lean
to storage
(tubs)

7½' x 2½'

Floor space 5' x 10"

Fold down Table
25"x29" x23" WLH

No shade
Bee escape
window

1/2 doors swing
both ways.

Brick
Smoker stand

Table

Bee house Shelter Drawing 2024

Bee tongue 2020

Chapter 5
WORKING THE AZ HIVE

Bee Seasons

With the AZ hive in mind

Bees function according to the environment around them, especially the seasons. Most beekeepers start the season with spring as that is when most the bees start to work. Bees get the hive ready for future queens and future colonies. In the summer the bees raise new bees and collect food for winter. Fall arrives for the last-minute winter preparations of storing food and creating winter bees. Winter is a quiet time but not without activity for the bees, such as keeping warm. Each season has things for the beekeeper to consider in order to assist their bees.

Depending on where you live the activity will be different. Here in the Pacific Northwest(PNW), United States, we tend to have quite a bit of rain all fall, winter and spring with some sun in the summer. But even within the PNW there are many differences in climate as well as the environment. West of the Cascade mountains is rain in the winter, while on the east side they get a little snow and hotter summers. If in the hills or mountains you get a lot of snow. Some areas have a lot of trees (west side), some have fields of grain (east side) and mountains in between. Look at your area, as to the climate, in how it may affect your choices. The information provided here is a guide to be adjusted as wished or needed. I do not profess

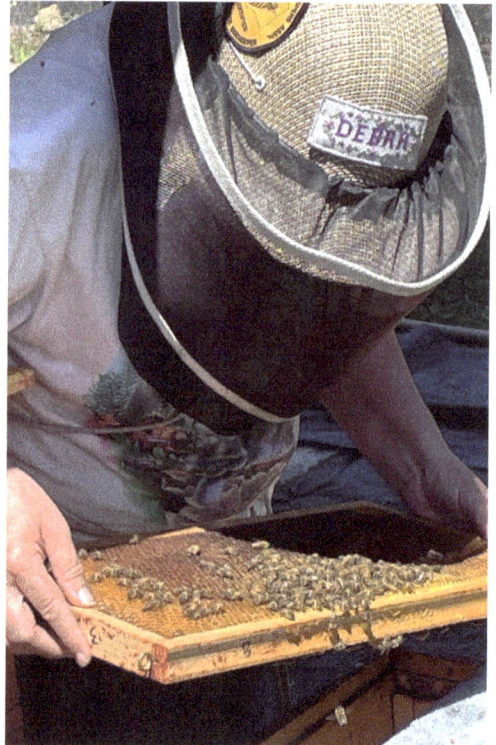

Debra examine frame
unknown Beekeeper 2021

Pollen Bee 2019

to know everything about bees, hives, AZ hives or even have a lot of information, I just have some experience and training. Ask around your local area, especially long-time beekeepers and read. But most of all use your brain to analyze your situation. Then chose as best as you can. I've listed the seasons with each one having general topics (AZ hive/house, food, pest, chores for beekeeper). I will be addressing these areas with the AZ hive in mind. I have no doubt I've left things out and that there are other opinions. Adjust your apiary as you wish. I have found most of the things a beekeeper does for AZ hives is the same as any hive of bees. If you already have bees, you will see most

Queen cell 2021

of this is the same as a Langstroth hive. Bees do behave like bees no matter where they live.

I have AZ hives in a shelter on the west side of Washington state. I live in the woods with a high amount of annual rain (130 inches or so). Food and weather are prime concerns for my bees in my location. Weather was one of the main reasons I chose the AZ hive.

Spring pollen 2019

Spring

(March, April, May)

Spring is the time of growth for the beehive. The hive goes into high-speed collecting pollen and nectar to build up the population of the hive. With more bees they can collect more and have more bees. This then creates the risk of swarming to create more beehives. This all creates some work for the beekeeper. The AZ hive has a little extra moving of frames, especially in the spring.

AZ hive/house

Sun hitting the front of the hive lets the bees know weather is nice. They then come out and search for food. Morning is a good time for the sun to hit the hive front (locate hive appropriately). As it warms up outside it warms up inside. The bees then can break their cluster no longer needing to keep heat. The queen then can begin to lay more eggs as bees can care for more brood. More space is then needed on the frames for more food and more brood. Bees build more comb to accommodate this. Drones are then needed to mate with the new queens about to be created (I add drone frames in the spring). All the space in the AZ hive maybe getting full. This then helps promote possible springtime swarming.

Empty Space 2018

The beekeeper needs to make sure their bees can build up, but not swarm. The big chore in spring is to make sure the bees have room to continue their growth. The AZ hive is a fixed space. If not tended it can be a problem. The beekeeper needs to move frames not just add another box as done with a Langstroth hive. Pull capped honey and replace with empty frames. Also add new frames for more brood (the queen needs space to lay eggs). This need for space is why I did the four chambers. Many in the U.S. do three chambers while in other countries they tend to have two chambers.

I have also noticed that the inside of the hive does not warm up as fast as the Langstroth hives. This makes the location more important so bees can start in the spring. I also chose bees that do better in the colder climates. I have Saskatraz or Saskatraz mixed with Carniolans. Perhaps a heated building could help this issue.

Food

Since bees are building up brood, they need pollen and nectar. Most places have local plants providing pollen and nectar. When you see pollen coming in it is the time the bees will start adding more brood. If food is not available, the beekeeper needs to provide it. Often providing pollen patties in the winter month of February or even March will help promote egg laying for spring bees. What you do depends on what your plans are for queens and building up the hive. I add pollen patties and sugar water (1:1) if needed. I feed sugar water in a feeder between the inner and outer doors located near the bottom of the hive. The feeder has screened holes for a jar to sit on top. Make sure the bees can reach the sugar water through the screen and the lid (lid has small holes). I set the pollen inside the hive on the bottom screen or on top of frames. Ideally, they have enough left from winter and have a food source in the surrounding area. Remove winter sugar early spring (they will toss out).

Pests

Spring is when the pests also build up. Yellow jackets are just starting to have their new queens out. This is a great time to trap new wasp queens in the area before their hives grow.
Checking for Varroa mites is needed. I start monthly mite rolls when weather is nice. I also do a single set of Oxalic Acid treatments early spring. (Set is 3-4 treatments 5-7 days apart) If dead mite counts are too high, I continue treatments until mites are gone. Mice are looking for a place to raise their young, do not let your bee house be that place. I check on a regular basis to remove them. I also add bad smells to deter them (lavender is one).

Chores

Spring is a busy time for the bees and beekeepers. Check for mites. Make sure bees have food. But most of all watch for swarming behavior. Losing your bees to swarming is not helpful to the beekeeper. Put out swarm traps. Check the hive often for overcrowding, drone build up and queen cells. Your best bet is to prevent your bees from swarming early. Make splits or move the queen (fake swarm) if needed. AZ hives moving frames around more than most Langstroth hives. With a fixed number of chambers this becomes important. I do a bit of "checker boarding" movement. I add new frames to build on, move full frames up (or down) or remove full frames. What each beekeeper does depends on what is seen and what they want the end goal to be. I also add drone frames (half foundation, half empty) to give them a place besides between my chambers to build drone comb. I need drones for mating new queens. Check your hives often in the spring, bees are sneaky and fast at swarm build up by making queen cells.

I like to keep good records of my hive. I record what is in each chamber and frame as well as what bees are doing. I have often gone back to my notes to remind myself what I did see. I start weekly checks of hives in the spring or even more often if I see things progressing in a direction I do not like, such as swarming. This is also time to plan for any grafting or splits. I start grafting when the hive has built up enough to support it.

Spring cleaning is also needed. Remove extra insulation, quilt box (early spring they will fill space with comb), and extra sugar. Make sure the hive is clean (bees clean some things but if you help, they have less to do – much like fall dead bees). I do not spend a lot of time cleaning the hive. I do clean the beehouse. As part of making sure all is ready for spring and summer use.

Always keep an eye out for disease. Treat as needed.

Summer bees 2023

Summer

(June, July, August)

Summer ends the bees build up and starts the preparations for winter. In early summer they will still have swarm potential but, hopefully, by late summer they are getting ready for winter. The beekeeper is busy making sure the bees don't swarm and have what they need to survive the winter. The AZ bee house is a great place to work in the heat of the summer.

AZ hive/house

The location of the beehouse in the summer is as important as in the winter but for the opposite reason. Summer heat can make hives too hot. The overhang on the front of the building should offer shade in the summer and allow sun heat in the winter when the sun is lower on the horizon. I have a screen on the bottom of my hives with a board under the screen. If it gets really hot I can add extra air flow by pulling out the board. Having a beehouse also gives the beekeeper a place to work out of the hot sun. It can get dark in the beehouse so add lights if you can. Or make a workspace outside for better light on projects. I've added a lean to on the side of the shelter for a covered outside work area.

Corn pollen 2022

The fixed space for frames means the summer is time to check that all frames are not full. If space is needed pull capped honey and extract or store. Move frames so there is space for new brood or nectar as needed.

Food

By the end of summer, it is dearth time. Often little natural food is available for the bees to collect. I begin feeding a ratio of 1 to 1 (equal sugar and water amounts). I feed sugar water and pollen as needed. New hives really need the beekeeper to provide food unless natural food is vary abundant. Near the end of summer, it is time to switch sugar water to a 2 to 1 (sugar-water) ratio. All my sugar water is fed inside (between inner and outer door in the feeder).

1 to 1 sugar water is building up the hive's comb and food. The 1 to 2 (thicker) is to feed and build up food stores for winter. It is dryer so takes less time drying before capping.

Pests

Summer is full of potential pests to a beehive, from ants and wasps, to mites and robber bees. A strong hive will deter most of these pests.

The yellow jacket will be in full killing mode by late summer. I have traps out, although strong hives tend to deter them. Robber screens confuse them, and small entrances allow the bees a smaller place to protect.

When forage gets scarce, large beehives also like to rob the smaller hives or just because they can. Do not feed a small hive at the entrance and close the entrance to one small hole (one bee at a time) and add robber screens.

Varroa mites continue to be a potential problem. Regular checks should be made to make sure mites have not taken over. I use a powder sugar roll, but there are several great methods. Treat as needed.

Ants tend to invade anyplace they can. I find more ants trying to nest in the summer. I have yet had success in completely eliminating ants. But, I have found they do not like cinnamon (I pour around the outside and inside of my shelter). I have tried several other things with limited success. I will continue to try new ant removal methods.

Wax moths are another potential pest. I try to keep an eye out for them and if I think I have them on a frame, I remove and freeze the frame.

Continue to watch for other furry pests such as mice or big ones like bears. For mice I use smells to deter, traps and removal. With bears an electric fence around the building is the answer. There are other pests, but I have not had them attack my hives yet (skunks, raccoons …).

Chores

Chores for summer include weekly or biweekly hive checks. I keep detailed records each time I check the hive. I watch for the hive to get too full as swarming can still happen. Remove honey frames as they cap unless you are leaving that frame for winter food. I store removed frames in a plastic tub. I start counting honey frames at the end of the summer (for winter food or my extraction). Make sure the queen is producing. Watch for diseases. Feed bees as needed. Make sure your bees are protected from pests. Add robber screens, especially to small hives.

Any time there is a dearth, with no natural food for the bees, the beekeeper needs to check the food stores and feed as needed.

Always keep an eye out for disease. Treat as needed. Summer is when I have seen problems, disease is not limited to summer.

Fall

(September, October, November)

Fall is a busy time for bees and the beekeeper, much like all seasons. Preparation for winter should be done by early fall. The bees gather the last of the food for winter. The beekeeper checks food, mites, winter spacing and any insulation for beehive or house. By late fall hopefully all is well and done for winter.

AZ hive/house

Preparation for winter means moving some things around. I have four chambers in my main hives and two chambers for queen test hives. I like the top chamber to have a quilt box for moisture and extra sugar if needed. All the top chamber frames are removed and stored in plastic tubs for the winter. Depending on the number of bees in my hive I may close the top chamber(s) and use the third (second) chamber for quilt box and sugar.

Depending on what the bees have done I may move the brood to the bottom chamber unless they are already there. I put honey above the brood unless the honey is already there. If I have a small hive of bees, I may move them to one chamber. They then would not have as much brood or honey. I may choose to leave them in two chambers but have empty comb or extra honey frames on the edges (I have 10 frame chambers).

The goal is to have the bees ready for a long cold/wet winter. I condense bees in the hive to only the space they need. Excess space does not help and may hurt.

I can have two hives in the four chambers with a queen excluder or solid board in the middle. Both hives need to be medium size clusters and have enough food for winter to use a queen excluder.

Once the hive is set for winter, I turn to the space between the inner and outer doors. I add insulation (open cell foam). I make sure there is air flow to carry the moisture away. I open any vents in outer door. I need air to come in the hive low (small front entrance and a small amount with screen bottom and board). I make sure the vents at the back outer door are open. Air flows through hive (bottom to top) and carries the moisture through the inner screen doors to the foam and out the back outer doors vents. If I do not have air flow, I have a lot of moisture at top of hive and between inner and outer doors. Moisture is needed for bees but drenching the hive wood creates warping.

Fall bees 2021

Fall queen 2020

Food

Make sure your bees have food for the winter. I leave 10 honey frames for each hive. I feed 2:1 sugar water as needed until about the middle of September. I add sugar (solid blocks or poured) at the top of the hive. I stop feeding the pollen patty.

This is the time I collect whatever honey I will get. I only collect excess honey, and not any sugar water honey. When moving frames around I remove my honey to a closed tub. I then take it to extraction site. I return empty frames to hive for bees to clean.

Pests

Varroa mite numbers explode as old bees die and only the winter bees are left. Mice look for places to live the winter in comfort (beehives are very comfortable). Then there are the normal year-round pests, like bears (electric fence works best).

Mites need to be treated and killed. Bees cannot survive winter when they are being eaten and given disease by mites all winter. Varroa mite numbers may explode in the fall. As bees die off, or hives die, the mites move to the bees left in the hive(s). There are many good ways to treat for mites. I recommend treating till the dead mite count is 0. I use Oxalic Acid vapor. I have a space at the bottom of the hive where the wand fits (got a Pro-vape the next year and made a hole in hive). I remove the divider board above vaporizer (can burn board), seal the inner doors (have a board that fits each door), close the front entrance so the vapor gets to the whole hive. In a few days I check under the bottom screen for dead mites. I have screen at bottom of hive (vaporizer sits on it) with a board under the screen closing the hive. Dead mites fall through screen to board under it. Sometimes I have had to treat every 5 days for long periods. Mice are the other winter pest that likes to invade beehives and house. I have a mouse proof entrance on year-round and have not had mice in my hive. But they do like to come into my shelter (everything is in plastic tubs). I remove nesting, add bad smells (lavender and baby powder (old left-over jar) to the storage shelves.

Chores

Fall is time to get set up for winter. I do moisture control with a quilt box and ventilation. It is time to close off chambers that are not needed. Move any frames you wish moved now as it will be too late later. I add the insulation between inner and outer doors. Remove all feeder equipment and clean.

Start cleaning any dead bees out. Do fall bee house and hive cleaning. I often do a real hive check (early fall or late summer) and take everything out to make sure I do not have any problems for winter. I clean up all my tools and put in a safe winter location (many things go in tubs for winter).

Always keep an eye out for disease. Treat as needed.

Winter

(December, January, February)

Winter is the time many think nothing is happening. But in reality, it is an important time for your bees. There are things the beekeeper should consider coming into the winter and check in the middle of the winter. The AZ hive system (hive plus bee house) makes it easier for the beekeeper and give the bees some winter protection.

AZ hive/house

The cold/wet weather means the bees cannot leave the hive as much. They do need to go out to "go to the bathroom". If it is too cold the bees stay in the hive to keep the queen, any brood, and them warm via clustering. Late in the winter they begin to gather pollen from early blooming plants. Heat retention and warm weather reaching them are important for survival.

The AZ hive and house provide good protection from the elements of winter weather. The insulation of having a building or shelter can be a game changer for beehives in the winter. But can also have negative effects. Having sun hit the front of the hive is important. It warms the front of the hive, so they know it is a nice day to come out. Using traditional AZ beehouse design (front roof arch) can help with a bit of heat retention. The way the sun hits the hive and how the air flows at the front of the hives affects the hive heat and heat retention. The AZ hive/house can better hold heat by not allowing wind chill removing the heat. But it may also keep the hive cooler in the winter because of insulation keeping the sun from warming it. Decide if you need insulation on the front of the hive (some places do not need insulation). Decide if you provide a bit of heat in your bee house (if you can). Heating the beehouse could help them hold heat but too much heat in winter makes the bees think it is nice out and one does not want bees out in a snowstorm. Heat should not be more than about 40 degrees (U.S.).

Winter snow 2021

Bees create a ton of moisture. If you have a wet space add more air flow. I have found it wet between the inner and outer door. I now have more air flow and added open cell foam in that space. I have more air coming in the bottom front of the hive and more space for moisture to leave the hive at the top back (inside shelter). You are not trying to dry the whole hive out but rather just keeping it from dripping on bees and warping the hive with too much moisture. In some dryer climates moisture may not be as severe as a problem.

Food

Winter inside hive 2019

The Bees have tried to store enough honey for winter but cannot always do that. This is especially true if the beekeeper extracts too much honey over the summer. I like to make sure they have 10 frames of honey/pollen for winter. Since bees like to travel up in the hive. I make sure there is food above the cluster. Usually this has been fine. I then add sugar at the top as extra insurance, so they do not go hungry.

Sugar can be added in several ways. I sometimes put newspaper down above frames and pour sugar on it. Or I make solid sugar cakes. Just add some water to sugar in any shape (I use tin pie pan) and it dries hard. I have also made frames of sugar (just pour sugar on foundation and spray water on the sugar it will then dry hard). Some add such things as a few drops of apple cider vinegar (keeps from molding) or peppermint or other types of nutrition (personal decision). Bees eat more in a warm winter as they are more active. Some winters 10 frames of honey do not do it. In addition, some hives have a bigger cluster, needing more food. I have also added several holes in my plastic foundation. This is to assist bees with the ability to move easier in the hive. When they break some of their cluster they need to access food elsewhere in the hive.

Pests

There are a variety of pests to keep in mind even in the cold winter weather. Hives and beehouses are shelter and food to various pests. Varroa mite numbers can explode as old bees die and only the winter bees are left. The mites will feed on your winter bees to weaken them for disease. I continue to treat mites till they are gone. Treatment can go well into the winter. Mice continue to look for places to live in the winter (beehives and houses are very comfortable). Then there are still the normal year-round pests, like bears (electric fence works best) or raccoons and skunks. Keep an eye on your hives and houses for an intruder.

Chores

Winter is a time when you do not need to check the bees that often. I check at least every few weeks or when it is nice out. On nice days you should look to see if the bees are out flying. Check food supplies. Make sure bees have food or they will starve. Some years they eat more or less than normal depending on weather and size of cluster. Add food as needed. Look for mice or other pest problems. Make sure there is not excess moisture in the hive. I do a general hive check – do I have live bees? Where is the cluster? Size of cluster? I do this hive check via peeking through the inner screen door (advantage of AZ hive). I do not open the hive. If it is cold and I must open the hive I heat the shelter space, so I do not cool the cluster. Early in the winter, summer bees die off, this usually leaves a great many dead bees at the bottom of the hive (the workers cannot remove them as easily because of weather and numbers of dead bees vs. workers). I remove dead bees and make sure the bees have their entrance clear for nice days. Some of these things may make the difference between surviving in the winter or not.

Dana working hive 2018

Entrance 2018

Inspection and Working the AZ Hive

Working an AZ hive is much like working any hive. Bees tend to be bees and do what they do. You will get propolis and bur comb wherever you do not want it. You must tend the bees. Unstick and scrape frames on regular basis. I add drone frame in spring (helps with bur comb).

AZ hives are similar to the Langstroth hives in that they both use chambers or boxes in a stack, have 3/8 bee space, use similar frames and usually have 8 to 10 frames to a chamber/box. Working the individual frames is the same for either hive.

AZ hives are different from the Langstroth hives in that entry is from the back of the hive not the top. When the chambers are full AZ hives are either split or removal of the capped honey frames replacing with empty ones to add space. They do not add new chambers/boxes to the top like a Langstroth hive. No lifting boxes-chambers as AZ hives chambers are permanent. Frames slide in and out horizontally instead of lifting frames up and out. The AZ hives are kept in some type of shelter or building with an outer and inner doors on the hive for access.

Things to think about or decide:

▸ Always work to protect the queen. Slow, steady and keep an eye out for her.

- Queens are often with the brood but don't count on it. Mine has been everywhere from walls to upper frames or even falling on the floor (reason for the hive table).

Empty center 2019

▶ Smoke or no smoke. Remember you are in a building – smoke will be with you. Smoker or smoke stick?

- If using a smoker, I keep it outside the building except when smoking.
- I put the smoke stick in a holder on hive table close to frames. I use ½ stick at a time.
- I use smoke mostly to get bees back into the hive. The Saskataz tend to come watch me and not go into the hive if I don't smoke them.

Frame spacer 2020

Frame Stand 2022 by Sherry Purkett

Hive Table Debra by Sherry Purkett 2022

- ▶ Do I need to move frames? The general hive arrangement is honey outside with brood in center.

Hive check procedures: (I always do those marked with an *)
- ▶ *Check the entrance – bees? Pollen? Fighting?
- ▶ *Check bottom board – mites? What has been dropped through screen?

*Bottom chamber check under divider board (entrance here) - dead bees?
- ▶ Use hive table if you wish.
- ▶ Open back inner door to chamber you wish to check.
- ▶ Unstick divider board -slide out and in carefully -use hive tool as needed.
- ▶ Unstick frames -hive tool – remove last frame for space. Slide to side.
- ▶ *Remove frames slowly, use rail and your hand to keep from dropping.
- ▶ Take one frame at a time out, check (honey, pollen, brood, larva, drone cells? Queen cells? Swarm cell? queen? pattern? Disease?) put back or put on frame holder.
- ▶ *To put frame back carefully/slowly slide in on rails and get in frame spacer slots at back of hive (flashlight might be needed)
- ▶ Check all frames and replace them. Can pull partially out or just peek in.
- ▶ Replace inner door. Use frame spacer to get frames set for door.
- ▶ Continue to check all chambers you want to check.
- ▶ Decide if you need to open the next chamber at top, checkerboard, remove honey, split, or treat.

Divider boards can be solid if you wish so no bees can pass, slotted for easy passage or queen excluder.

Always keep divider boards in the hive with bees. Otherwise, they will fill excess space with comb.

I empty the whole chamber if I wish to clean chamber walls or isolate the one above or below. Or just for ease of checking frames.
Quick check – look in though screen of inner door with flashlight.

Feeding:
- ▶ Put feeder board in back of hive at bottom.
- ▶ Add or change jars of sugar water. Open only the outer door.
- ▶ Add pollen paddy on top of top frames or on bottom.
- ▶ Add sugar in the empty chamber at top. Put on newspaper if not using sugarcakes.

Treating for Varroa mites: Follow directions of whatever method you choose.

Hive Inspection Sheet

This is my AZ Hive Inspection sheet. I designed it to fit me and my hives. I have tweaked it several times and expect to continue tweaking. I looked at many samples of what others are doing, worked hives and then decided what I wanted to document. Each time I go into the hive I have a plan (it changes sometimes when I see what is happening). I fill out parts of the sheet as needed. This often gives me information for next time or a yearly record of reoccurring events. I encourage you to decide how you want to document what is happening with your bees. You can make a notebook like this, have a tablet, use taped notes on the hive, have blackboard on the inside of the outer door (I have) or do an on-line program. Do what fits you and your hives.

For instance:

▶ In the late summer, I do a honey count to make sure my bees have winter food. I normally write a bit more detail about honey on each frame than in the middle of the summer. Then in the spring I can go back and see what the bees really ate. This can be valuable information for winter and next year.

▶ Things like Varroa mite treatments are just put on the back of the current sheet. For mites I write the date treated, number of dead mites for that treatment and total dead mites. I get dead mite count off my bottom board as dead mites drop through the screen. I check just before treating (and sometimes a few days after because I'm curious).

▶ Grafting new queens – I look at the last inspection to see where my best larvae might be.

▶ How much sugar water or pollen are they using? Tells me the needs of the colony.

▶ Things like weather, time of day may determine if I need smoke or not.

▶ Movement of the frames. My frames are all numbered so I can see where they have been.

▶ I also make random notes – such as: Name of the queen. Things I want to do next check. Adding a queen excluder at a location I have added.

A sample sheet follows: I've removed some excess space and repeats. I use a full sheet in the summer, with 3 to a page in the winter. The sheet for a winter hive that is not normally opened has less information. I've put some notes in Italic *Hive names* – Yellow, Blue (I use very creative names for my hives).

HIVE INSPECTION; *hive names* YELLOW   BLUE   **Weather** conditions:

Date/time   **Queen;**   seen?   *Where*   *Marked?*   *color*

Bees- Calm   agitated   aggressive   *smoke used?*

Hive condition: strong   ok   weak   propolis   burr comb   disease?

Bees out? *This is my entrance check. I write what they are doing at entrance if needed.*

Chamber #? 2 - **Frames;** <u>Honey</u>/nectar/   Pollen   <u>Brood</u>   capped?

<u>Cells</u> -queen   drone   pattern   eggs   larva

I add details as needed for each frame. Each box represents a frame. I put frame number at top.

Chamber # 1 - Frames; Honey/nectar/   Pollen   Brood   capped?

Cells -queen   drone   pattern   eggs   larva

Action taken-<u>food</u>? Sugar-Water   sugar   pollen-paddy   <u>Moved frames?</u>

Bottom Board: *This is my hive activity and mite drop check – you can learn a lot by what is on the bottom board under the screen.*

       WINTER HIVE INSPECTION;   YELLOW   BLUE

Weather conditions: **Date:**

Bees- out?   Calm   agitated   aggressive   **Queen;** seen?

Hive condition: strong   ok   weak   propolis / burr comb   disease?

Size of cluster   Dead bees?

Chamber 1, 2, 3, 4 - **Frames;** <u>Honey</u>/nectar/pollen capped? <u>Brood</u>   capped?

Location of cluster_____   <u>Cells</u> -queen   drone   pattern   eggs   larva

Action taken-<u>food</u>? sugar   pollen-paddy   <u>Moved frames?</u>   Quilt box? Treat?

Issues & Solutions

(Swarms, robbing, mites, drone or excess comb, moving, splits/queens)

Beekeeping has many things going on besides the regular work of keeping bees (food, checks ….) and sometimes there is a problem. This is a look at a few issues and some potential solutions with AZ hives. Please note that it is very similar to what one does with any hive. Also note other topics such as "Pests and the AZ Hive" and "Bee Seasons" have some of the same type of information. Remember these are just some ideas and there are many more ideas out there.

▶ **Swarms**

The beekeeper needs to make sure their bees can build up but not swarm. The big chore in spring is to make sure the bees have room to continue their growth. The AZ hive is a fixed space. If not tended it can be a problem. This need for space is why I did the four chambers, but it can be managed with two or three chambers as well.

- *What you may see:* Every or most frames full of bees, brood, nectar/pollen or capped honey. Queen cells at bottom of frames. Lots of drone cells or drones. It is spring and there is plenty of food.
- *What can you do:* Move frames not just add another box as is done with a Langstroth hive. Open the next chamber up, if you have one. Pull capped honey and replace with empty frames. Also add new frames for more brood (queen needs space to lay eggs). Flip frames front to back if there is space on one side. They will build on the other side. Add swarm traps or scions for your swarm to go to. Check weekly (or more as needed). Do a split. Remove some of the hive to another hive. You can return later. This could include the queen.

▶ **Robbing**

Robbing can completely kill a hive in one day. The robbers can be another honeybee hive or wasps. Both will come to steal honey. Wasps also want larvae for protein. This is common at the end of the summer when food is scarce. Robbers will return for days until they see there is nothing left to steal.

What you may see: Excess bees fighting and trying to get into the hive (including any holes or cracks). Dead bees at the entrance. Junk at entrance (wax pieces or honey trail). Wasps coming in the hive. Inside the hive- wasps, frames with capped honey torn open (ragged edges). More fighting and dead bees. A frenzy of activity.

What can you do: Before *robbing*: Keep a strong hive. They have a chance to protect it. For weaker hives -_Remove the feeders (especially at the entrance) during the potential robbing season. Close down the entrance to one bee wide. Add robber screens (confuses robbers as to where entrance is), trap wasps in early spring (gets queens), using wasp traps. *After robbing starts:* Close hive, move hive a mile or more away and add robbing screen. If you can hang wet towel at entrance or have sprinkler it will also help remove them for awhile.

▸ **Varroa Mites** (see Chapter 5 "Varroa Mites the AZ Hive", "Pests & the AZ Hive")

- *What may you see*: mites on bees, dead mites at bottom, health issues
- *What can you do:* Check for mites on a regular basis and treat as needed.

  Varroa mites are a problem for most beekeepers. There are many ways to treat for mites. I use Oxalic Acid vaper with a wand or vaporizer. I only have a few hives so do not need the latest, expensive equipment. I treat in the spring and fall (till mites are gone). Generally doing a set of 3 treatments a week apart. I can check the board under my hives for a mite count. I have screen bottoms with board under my hives. This gives me a real idea of how many mites were in the hive.

▸ **Drone comb or excess bur comb**

  Bees love to build comb and drone comb in places we do not want. Drone comb is a different size than the foundation we purchase so they make what they want. Space to lay eggs or put in nectar if needed and if you do not provide space, they will make their own place. No divider board or frame – bees will fill with drone. Keep excess space to minimum.

  - *What you may see*: Bur comb someplace it does not belong. Usually, they build comb around the frames (bottom, top or between chambers on divider board). Or any place you do not have a frame but should.
  - *What can you do:* Remove bur comb. If only a little bit you can just slide the front of the frame to the side to disconnect. If there is more, then use hive tool to cut bur comb. Add empty frames as needed. Add drone frames (half foundation, half empty) to give them a place to build drone comb. Don't use foundation.

Part plastic foundation. Top 1/3rd to 1/2 for foundation with bar added on the bottom side to hold foundation. Or you can put the bar going down. Or add 1/3 plastic foundation strip in center leaving both sides open. Drone comb is commonly built on the bottom empty half. Nice way to allow for plenty of Drone cells and they fill with honey for the winter.

▸ **Moving or Splits**

Sometimes we need to move our hives. Unless you put your AZ hives on a trailer or have not connected them to the beehouse, it will be different than picking up a box.

- *What you may see*: You want or need to move your hive or part of it. Robbing causes a need to move, split, moving to new place, selling hive.....
- *What can you do:* If just moving in the bee house – just pull the frame you want and move to a new hive. Close up both hives, add something to the entrance such as a branch or robber screen. They then need to reorient. Open in a day. Forager bees will most likely return to the old hive if it is open. Move some distance (a mile or more). Transport bees in a nuc type box with spacers to keep the frames set. If making a nuc for things like grafting or split move bees to a nuc. Nurse bees will stay with the brood. Many bees will also tend to stay with the queen and brood.

Drone Frame 2020

▶ **Moving AZ to Langstroth or Langstroth to AZ**

For short periods of time, you can put the wrong frame in either Langstroth or AZ hives. Frames are designed to fit and work well in their hive. The Langstroth frame just fits into the AZ hive. It will collect bur comb and propolis making it hard to remove because the frame top and bottom are not concave. The AZ frame needs support to fit into a Langstroth hive. There are plastic ones on the market, but I found a simple screw or eye bolt placed on the top side works. This is to hold the AZ frame in place of the frame on a Langstroth frame. You must space them carefully.

Combining AZ and Langstroth frames in one hive is a good method if you are changing from one to the other. Add new style frames and allow bees to fill then remove to appropriate hive.

Pests and the AZ Hive

The AZ hive is no different than any beehive, they have potential pest problems that the beekeeper needs to tend. There are beekeeping books and a ton of information on the internet about pests and how to treat them. Treatment is not much different for the AZ hive.

The challenging pest year for me, I started with great hives with bees building up wonderfully. I did my normal mite treatment, and I had 0 mites – fantastic! The swarm season started, and I addressed it. I began my grafting to get my next set of queens. I naturally used my best queen. Which was a Saskatraz mix from my original queen. I wanted to get as many new queens off her as possible. She seemed to keep mites off, built up nicely, was sweet and made honey. Things were going well. Then the fun or nightmare started: a dearth at the wrong time of year due to bad weather, robbers, ants, more robbing, yellow jackets and bald-faced hornets, wax moths, mice/rats and mites. I handled each as they came wave after wave over the summer. I was wondering if I would have any bees starting fall. These challenges can come and go regardless of the style hive you have.

There are some general things to do for potential pests. Feeding when needed (nucs usually need feeding). Keeping entrances open only as needed (nucs always need small openings). Checking for mites, ants, wasps …. But most of all keeping a healthy and strong hive.

▸ **Ants**- Ants seem to be a problem for most beekeepers. Keeping the area around bee house/shelter picked up helps. I live in the woods, so I always have ants. I move things in the shelter on a regular basis looking for ants and removing them. I found a big hive of ants on top of my hive in the insulation space. I found ants below my hives. I have closed up holes they have used to get in between my structures. I added at various times- under my hives in the shelter (not a place that has bees) cinnamon, Borax (with powder sugar to attract them), and Diatomaceous earth. All to make the spot not like their home. I expect the ant fight to continue. The best method seems to be cleaning them out, use deterrents and hives that do not have insulation voids for ants to enter. Instead use thick solid wood for the front.

▸ **Mice/rats** – I always have mice and rats (chipmunks this year too) coming into the shelter. They look for food and nesting places. I just keep them cleaned out using traps.

▸ **Varroa Mites** – Mites are a problem for most beekeepers. There are many ways to treat for mites. I use Oxalic Acid vaper with a wand or vaporizer. I only have a few hives so have not got the latest equipment. I find I have mites in the fall and few in spring or summer.

I treat until mites are gone. With the AZ hives I can, and do, treat in most weather. If it is cold, I heat the shelter a little, so the hive does not lose heat. (see also Chapter 5 "Mites the AZ Hive", "Issues & Solutions")

Varroa Mites 2020

▸ **Moths** – 2020 was my first year to have moths get into the hive. They had in the past stayed at the bottom where there were no frames or bees. With all the fun of moving bees and losing hives I had some frames and chambers without many bees. I found they had laid eggs and I had the webbing on frames with a few moth worms hatching. I cleaned them off and moved all frames to freezer (2 days or so) to kill any moths that were left. Forgot to take photos, sorry! Now I am not leaving frames with anything on them in a hive that has no bees. I am doing a freeze on the frames, then storing them in sealed plastic tubs.

Wax Moths 2021

- **Robber bees** - I live in a remote location and have almost no houses for miles around me, just forest land. I do have a few beekeepers within about a mile or so. Some take good care of bees; others get bees and lose them (no food or treatment). One day I was surprised to find a basketball size swarm attacking my nucs. I was at end of first grafting and starting second grafting session, so I had many nucs and only one big hive. The weather had created a dearth -I was feeding. All the hives had one bee entrance except the big one. I lost three hives (nucs) including my next queen and my grafting queen in a short period of time. I checked several items to determine where the robbers were coming from. Not my bees. I closed up my hives to check the direction bees came from and left. I added white dots to the robbers helping me identify them. I suspected they were hungry, and the opportunity arrived. For future robbers I'm going to feed them some colored sugar water to see where they are from. I removed nucs to another location a couple of miles away in the opposite direction they came from. I added robber screens to all my hives (even big colonies). I closed up my bee hives and dispatched the robbers. I had a return visit of the robbers and did a repeat of care (screens were already on). I did not lose any colonies during the second bout of robbing. After the first robbing I was not sure I would have any bees left. The remaining queens were due to be replaced. The robbers had torn through my bees (nucs) winter food stores and killed most of the bees. The robbing and killing happened in one afternoon.

  Keep tabs on your hives, especially small colonies to identify robbers. Wild acting bees, fighting bees and torn open honeycomb (bits on bottom board). Add robbing screens in late summer and small colonies. Do not feed outside of hives which can attract robbers. Do not wait for a sign of robbers to add the robber screens.

- **Wasps (Yellow jackets and Bald Faced hornets)** -Both show up every year. I trap early in the spring to kill queens and then later to keep their hive numbers down. I remove hives in the area that I find. I just use the traps with food in the bottom that wasps go in but can't get out. Killing queens early has made a huge difference. A fly swatter works nice when you see them on a surface near or on your hive. The key is to not let the wasp population get too large. They are part of nature and small numbers are good.

Wasps, Yellowjacket 2021

- **Bears** – While I have not had bears bothering my hives, I do know beekeepers that have lost hives to bears. One bear in a few hours can destroy an apiary. The bears usually return. The solution is an electric fence or full bear cage around your hives.

There are plenty of other pests I have not yet experienced such as raccoons, skunks or hive beetles and hope to avoid.

Varroa Mites in the AZ Hive
(Oxalic Acid vapor)

The AZ hive is no different than any beehive in that they have Varroa mites. There are books and a ton of information on the internet on mites and how to treat them. Treatment is not much different for the AZ hive. Each beekeeper needs to decide what method they wish to use. Note that the information regarding treatment methods is in constant flux.

Over the summer I check for mites with a powder sugar roll (not perfect but gives me an idea). In my area I find I have mites in the fall and few if any in spring or summer. I do a short treatment in the spring to start the off good for the summer. It is important to have mite numbers down, or gone, for winter bees. The development of winter bees is important. A strong bee does better in the winter, so healthy food and no mites make a difference. I treat them with Oxalic Acid Vapor till mites are gone. With the AZ hives I can and do treat in most weather. If it is cold, I heat the shelter a little, so the hive does not lose heat.

Bees with powder sugar 2018

Checking for mites:

Before treating you need to see if you have mites. There are several ways to do this. Sticky boards, powder sugar roll, liquid detergent, alcohol wash, mites in drone larvae or even counting dead mites on bottom board after treating. There is a lot of information out there on mites, both checking and treatment. Most are fine. Researchers tend to do the alcohol wash and count before and after treating. All methods are not 100 percent effective. Each does things a little differently. I expect more new ideas each year. Find the ones that fit and work for you. The Powder Sugar roll is one of the easiest ones. You can do a detailed sugar roll or just a quick one for mite count. I use it because I just want a quick check to see if the mites are out of control. I do not do the percentage. I know I will treat in fall and spring. I do check each month in the summer. If I have too many mites, I remove the honey frames and treat.

Oxalic Acid vapor

I use Oxalic Acid vaper. I treat with it in the spring and fall (till no dead mites drop). Treat at least 3 to 4 times as mites do not die under capped brood. I tend to do about 5 to 7 days apart to catch as many mites out of capped brood as possible. Treatment is best in warmer weather when bees are out. I treat twice a year or as needed (spring and fall). In the fall I do treatments 5 days apart. I check the board

Mite Wand tools 2020

under my hives for a dead mite count. I have screen bottoms with a board under my hives. This gives me real information of how many mites were in the hive that I have killed, and if I will likely still have to treat more. The AZ hive design makes a simple check for mites on the bottom board very simple. Like all hives, you must seal the hive before treatment to keep the vaper in during treatment. I have cut boards to fit over the inner screen doors. I tape or add foam as needed to seal most cracks.

Supplies: A vapor wand or vaporizer, power can be a 12 volt battery (AC converter for the vaporizer) or AC extension cord, Oxalic Acid (OA) powder (Wood Bleach), gloves (poison to touch), good gas mask with proper filters (the vapor is toxic to humans), small measuring spoon to scoop Oxalic Acid, a way to close up the hive and all openings, (use boards, tape, cloth, foam, etc. as all vapor needs to stay in hive), and a timer or clock.

Mite Vaporizer tools 2021

Warnings:

▸ Oxalic Acid vapor is not ok to breath or to get Oxalic Acid on your skin.

▸ Use only outside with air flow. Check wind direction away from you.

▸ Wand dish (Oxalic Acid holder) gets hot and could start a fire. Monitor and use in a fire safe place (mine is on a metal screen with any wood above moved out)

Directions:

Wand or Vaporizer

1. Do a test run of the wand or vaporizer to see timing with your power (different sized batteries give different times). Sublimation timing varies. Do in open air.
2. Remove your honey supers or frames.
3. Suit up. Mask, gloves and goggles.
4. Seal up hive (close front and any holes)
5. Put about ¼ teaspoon into the dish, cap or container per brood chamber. Amount has changed over the years – check current recommendations.

Mite Vaporizer tools 2021

Wand method:

6. Put the wand into bottom of the hive (seal)
   (remember it gets hot enough to start a fire)
7. Connect to the battery terminals
8. Move away from the area to avoid fumes.
9. Wait appropriate time (usually 2-5 minutes – mine takes 7 with my small battery).
10. Disconnect from power (battery or cord)
11. Open up hive and remove wand after a minute or two.
12. Cool wand before next use.
13. Wait 10 minutes (more or less) for the vapor to get all over the hive before reopening the hive.

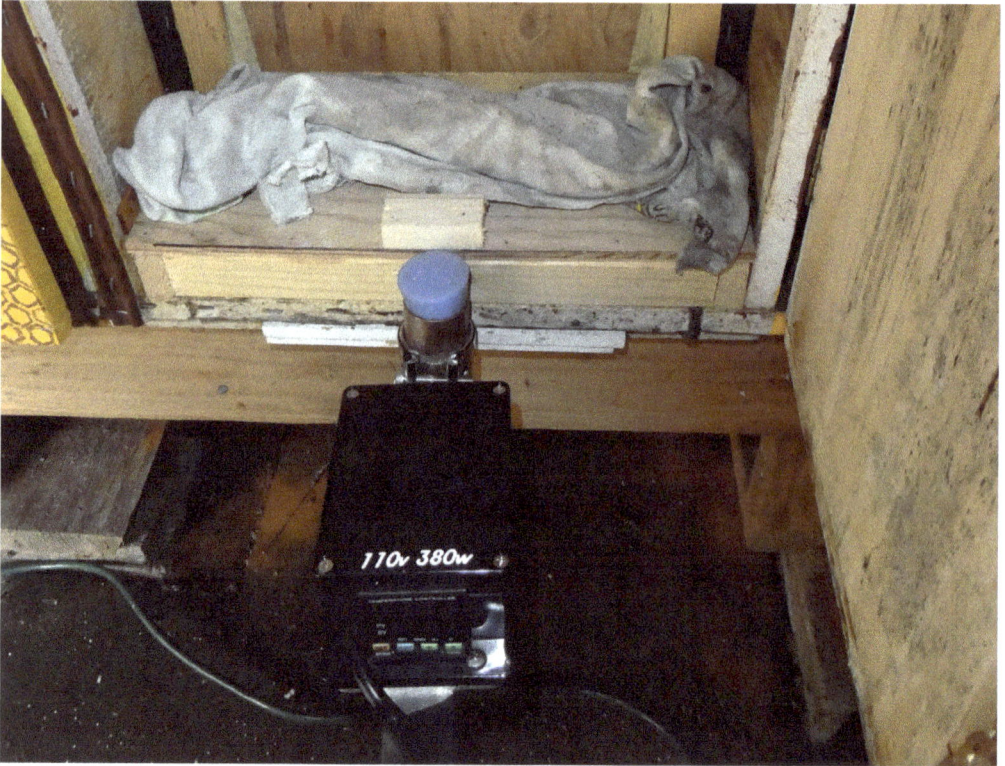

Mite Vaporizer tools 2021

OR

<u>Vaporizer method:</u>

1. Heat vaporizer up to 400°
2. Put vaporizer in hive before or after heating upside down so OA does not get in metal heating dish.
3. The cap holding OA needs to be flipped to add the OA to the heating dish after vaporizer reaches 400°.
4. Once flipped and OA is in the heating dish the temperature drops.
5. Let it get back to 400° (this is vaporizing temperature).
6. let it set for a bit before removing it.
7. Disconnect from power (battery or cord)
8. I clean with water between each hive as the spout sometimes gets clogged.
9. I do not need to wait to move to next hive. I start next heating after cleaning.
10. Wait 10 minutes (more or less) for the vapor to get all over the hive before reopening the hive.

Wand

I remove the bottom divider board (the wand can create a small fire) and sometimes move the frames above the wand. The wand sits on the bottom screen with the handle between the inner and outer doors in the bee shelter. I use a small 12-volt electric trolling motor battery for power.

Vaporizer

The vaporizer needs a small hole to enter the hive. Drill hole for the vaporizer spout to enter hive. This can be in the front or back of the hive. I made a holder for the back bottom space above screen bottom board. This is also my place for feeders. I cut a copper tube to fit over the spout long enough to reach the center of my hive. This allows the vapor to be in the same place as the wand. The other place I use is in the inner door. I drilled a hole in the board I placed over the screen. I insert the vaporizer spout next to the screen.

I plug the vaporizer into a battery charged by a solar power source for power to heat vaporizer. Different vaporizers use different power sources some use a battery and some AC power.

Mites also mentioned in Chapter 5 "Issues & Solutions", "Pests & the AZ Hive")

Queens in the AZ hive

New, moving or creating.

(see Chapter 2 Getting Started, "Adding Bees to AZ Hive" and "Frames-Langstroth to AZ")
The AZ beehive uses queens like all hives. Bees create new ones when their queen gets old, dies or they want to swarm. Beekeepers, like bees, want to create new queens or move them around too. This is a quick look at how that can happen in an AZ hive. Creating queens (grafts, split, removing queen cells), placing new queens in a hive, moving a queen, creating a new hive (split), or removing a queen (combine) will be explored. There are many books and articles that give details on how each of these are accomplished. I will only cover the part that connects with an AZ hive. Bees are bees and working with the queens is very similar in all hives.

Queen Cage;

Creating a new queen or just moving a queen is often done in a queen cage. The cage is to allow the new hive to accept the new queen or for her protection while moving. If it is her bees, just release her. If it is a new queen, the cage should be in the hive for a few days to assimilate the hive to the new queen pheromones. Either use a slow release (marshmallow or sugar plug) that the bees remove, or one that you remove after you are sure they accept her.

To leave the queen cage in the hive, remove one frame in the brood area. I have several ways of caging the queen in the hive. If I have a new queen in a Denton 3-hole cage, I hang or connect cage to a brood frame next to an open area created by removing a frame. Opening for the queen is pointed down so bees have access to feed queen through the screen, just like in a Langstroth hive. I also sometimes use a push-in cage if I want her to have more space to lay eggs. I have various

Queen Cage on frame 2021

Queen Screen 2019

sizes from small to half of the frame. A push in cage can be used over queen cells, to hold them until they emerge. Put a frame with the cage into the hive leaving the cage in the open space where you removed a frame.

Things to watch out for, or think about:
When putting the frame with the cage on it in or moving out of the hive, be careful not to bump the cage off.

▶ Make sure a push-in cage is pushed in on <u>all</u> sides (I have had escapees).

I find that rubber bands hold all cages well.

▶ Check - if the bees are tearing at the cage, - they are trying to kill the queen. If they are feeding her then they accept her. I watch for a few minutes to see that she is accepted upon release.

▶ Decide if you should leave some of her nurse bees with the queen in the cage or not. I've done both.

Combining colonies:

Combining hives is done when one has no queen or hive, or hives are too small to thrive or survive. If a hive has been queen-less for a while they often just enter the hive with no fighting. But if they think they have a queen and are being invaded you lose many bees to fighting. In the AZ hive you will be adding the new bees (usually) in the chamber above the one with the queen (often stronger one). I empty the chamber above the hive that is staying. Then I just put a piece of newspaper under the frames and rods of the top hive. I usually make a few small holes or cuts in the middle of the paper. Make sure the hive that is moving has no queen. Place the frames with the moving hive into the empty chamber. Wait a few days and check. If they are flowing up and down, remove the rest of the paper.

Hiving a colony:

A new colony is often welcome and easy to add to an AZ hive. Options vary on how you acquired the bees. If it is a swarm you captured, hopefully, you have used AZ frames. If not check out how to change frames in Chapter 2. The queen may or may not have been found. If in a cage release her into the AZ hive. The frames just go in. Or if you have a blob of bees, I like to put them in the chamber above the one I want them in (they just go down). If the queen is in the hive the bees can be dumped on the hive table. They will just walk in after the queen.

A new package is done much the same way. I keep queens in a cage on a frame for a few days. I dump the bees on top. Then I leave the package box upside down in the upper chamber. I remove the package box in the next day or so.

Package in hive 2021

Moving a colony from one hive to another is harder. The bees often want to return to the old hive. This is fine if you want bees there too. I just move all the frames to the new hive. I close it for a day or two. I continue to move bees if the hive needs to be empty. Bees like to stay with the brood and the queen. When you move bees, you can put a branch at the entrance to help them think about re-orienting. Another option is easier. Put the whole colony into a movable hive and take the hive a mile or two away for a week or so. Return them and place frames into the hive you want. Remove the travel hive. I have a 5-frame nucs for this.

Nucs: AZ nucs are designed to accommodate AZ frames.

AZ hives are often 10 frames per chamber. But there can be a need for smaller 5 frame hives. This can be done by having 5 frame chambers in your bee house and some standalone 5 frame nucs. Both are useful. This smaller space is used for smaller colonies. I use it for grafting, walk-a-way splits, separating a colony due to robbing or swarm behavior.

Debra Nuc 2020 unknown

Nuc inside 2020

In the spring when the colony is growing, they think about swarming. I find this is the perfect time to remove some of the brood for grafting or a walk-a-way split. Or if the hive is thinking of swarming, I can remove the queen with a bit of brood. I can then return her later. This usually settles them down.

The 5 frame nuc is standalone. The design I'm using was created by Dana Schack and then tweaked by me. It is much like the Langstroth in that the frames are added in and out from the top. It has a divider in the middle and bottom to separate it into two two-frame hives that can be added. That allows 2 new queens. Entrance each end. Frame spacers are on the inside ends of the hive to keep bee space.

Honey in the AZ Hive

AZ hives produce wonderful thick honey. 12% moisture is not abnormal. The design of

Honey 2018

the hive and beehouse allows the bees to dry it a bit more than most hives before capping. Traditionally, in Slovenia, the honey frames are gathered as they are capped, a few at a time. Since they tend to have only two chambers, space is needed for more nectar or brood. This process consists of removing a capped honey frame and replacing it with empty frame. They then either extract honey or store it for future extraction. This also allows the extraction of different types of honey as the nectar flow changes.

Frames

Before gathering, the beekeeper needs to decide if they want to extract, crush and strain, or have comb honey. Each uses different types of frames or bars.

▸ Bees can make all their own comb with just an empty frame. I find comb using this method a little precarious so wiring the frame is advised.

▸ Plastic foundation can be inserted into the frame to start the bees and give additional strength to the comb.

▸ A wood bar in the center, length ways (or squares) can allow for cut comb.

▶ Part plastic foundation. Top 1/3rd to 1/2 for foundation with bar added on the bottom side to hold the foundation. This also works for drone frames.

Removal of honey frames

Honey gathering starts with removing honey frames from the hive. This is done to allow bees space for more nectar or brood. It is also done just for extraction or if closing that chamber. Use fully capped honey frames.

Removing the bees and protecting the honey frames from the bees is needed before extraction.

Brush bees off the frame and place frame in a tub or in a honey room (if you are in the beehouse). I like to brush off the bees on a sheet outside the bee entrance. Then place the frame in a container (I use plastic tubs) to keep bees off the frame. The bees are attracted back to the honey frames, so containment is necessary.

If using the chamber add a new frame to replace the honey frame.

Honey Extraction

Bees will follow the smell of honey. Bees will steal it unless you protect your honey from them. Remove the honey to a closed location to extract, crush or put it in containers.

A heated room works best to allow honey to flow. Expect a mess, removing honey with any of the processes can create a mess with honey all over. I let bees clean my extractor, honey press and empty frames (I put back into their hive). Commercial beekeepers, sideliner beekeepers and bee clubs often have a honey room. If you can, add a closed honey room with water into your beehouse.

Bee removal ramp, sheet 2018

Honey Frame 2023

Extractor

There are many sizes and styles of extractors, from electric to hand cranked, and two frames to multiple frames. They use centrifugal force to throw the honey out of the comb. There are a variety of tools to assist you. Have honey buckets ready. Work over a large tub.

Open the wax caps. They can be cut off or heat melted and scraped.

- Place in the extractor. Turn on and then when side one is empty, turn frames to get the second side. Remove and repeat till done.
- Pour honey from the extractor into a honey bucket using strainer to catch any particles. You can strain again if you need it cleaner.
- Bottle after letting it settle (I like overnight)
- You can return the frames to the hives for the bees to clean.

Crush and strain or Honey press

If your honey is not in a frame (like top bar) you can crush and strain by hand or use a honey press (fruit press).

To crush and strain, place in strainer and crush until in very tiny pieces the honey drains out into a bowl under. The wax needs to be strained out.

For a honey press, put the honey with comb in strainer bag (honey bag) then press. The honey drains out all strained.

Bottle or strain again if needed. You can process the wax after.

Comb

Use new comb for your comb honey. Cut pieces and place them into a container. Glass jars look nice with some honey poured in with the comb honey.

Extract 2017

Honey pour 2018

Bee

Chapter 6
DESIGNS FOR THE AZ

AZ Hive Construction
Variations, modifications & likes

My first AZ hive was designed and constructed without ever seeing or using an AZ hive or obtaining plans for an AZ hive. There was very little information at that time available in the U.S. Since then, I have made several changes to my design and hives based on my use. Below are things I would change and some things I like and have kept with my hives. The designs drawings that follow are not full plans. They are drawings of what I have done and have had success using. You need to decide what options you wish for on your hive and what lumber and supplies are available, as well as exact measurements.

Changes
▶ Hive body:

- It is better to build one or two chambers and attach other chamber(s) to get to a 3-4 chamber hive. Thanks to Dana Schack for this design.
  Why? It controls warping walls better and it gives control on the number of chambers.
- I do not recommend a 2-chamber. I found it too small. The honey did not get capped, and it is hard to work from the top if it is full of bees.
- I did 4-chamber with the sides in one tall piece.

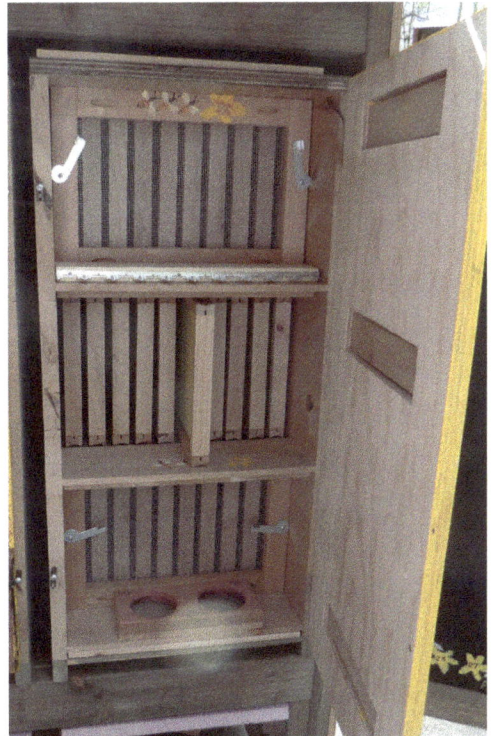

3 chamber AZ 2018

Roadside hives 2023

▶ Front of hive:

- Do not leave insulation space. If you want more insulation nail/glue two boards to get greater thickness. Or buy thicker wood. While the void I created was sealed it still got ants chewing into the foam void.
- Not all places need the insulation to help bees with warmth or cooling.
- If you have a temperate climate, I would not use insulation.
- I had air space with insulation on the front of the hives. I had ants in the void and insulation. I have since added solid wood in that space.

- ▶ **Bee entrance:**
  - • Upper chamber entrances. I would do the disk opening with hinged landing boards. I currently have openings with removable landing boards. Mason Bees are tending to use the holes that are meant for the landing board attachment.
  - • Bottom entrance. If one insets the hive slightly from the front of the bee house, there is a narrow ledge in front of the hive to serve as a landing board.
- ▶ **Rods/bars:**
  - • Rods work well but there are times it would be good to pull them out. Make a channel to slide in and out. This seemed too hard at the time I built. Or drill a hole at the side of hive to slide in. Leave a little extra rod to pull out. This only works with space to side of the hive or standalone nuc.
- ▶ **Divider boards:**
  - • I use metal L shape braces to hold and allow divider boards to slide in and out. But they must have the screws on the bottom side of L brace otherwise screws block sliding boards–
    I have half and half). Photo is of the <u>wrong</u> way.
  - • Braces also need to go the entire length of the wall as well as the back wall. This will help block cross chamber travel if you wish to close a chamber.
  - • I am experimenting with round holes instead of the long slots for the divider board. 1st year it went well. Size and spacing to be determined with more use.

Disk bee entrance & handle

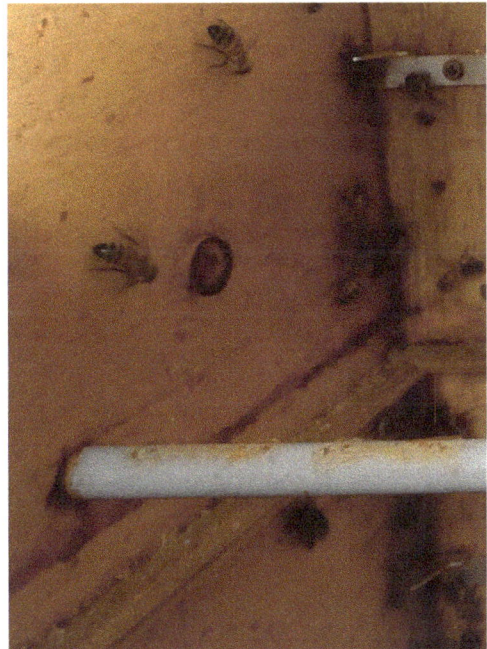

Divider board brace 2020

Inner door cloth 2020

Inner door support 2020

▶ **Inner hive doors:**
  - Hive back/ inside beehouse.
  - Do not make the fit too tight – I have too much swelling and shrinking of doors here in the Pacific Northwest (Washington State, USA). Add a strip of outdoor cloth around the top and sides. After trying several items, the cloth works to keep the bees in the hive and allow the door to come in and out easily. Tested on several hives for about 5 years – with wonderful results. The cloth is folded and stapled onto the inner door.

- ► **On Hive body:**
  - • I have added a permanent board under the inner door. Place at the same height as the rod. This is for the inner door to sit on. This can be an inch or several inches if you want to have feeders on the board. I have the middle chamber wider. It is easier to work the hive if you use 1 to 2 inches. It could have a slot to pull in and out for easier access to hive or different sizes of boards (for feeders or just board).
- ► **Outer hive door:**
  - • Make a separate door for every two chambers. This reduces the chance of warpage, better seal, and easier to lift if removing.
  - • Use lift off hinges. Thanks, Dana Schack, for this idea.
  - • Mount the door such that you can use space next to hive. I would put only two hives next to each other with space for a table on each side. Mount the outer doors so they open toward the center between hives. One right, one left door opening. Workspace located next to the open hive makes working hive easier. Or have a small table that you can move next to your chair.

Likes

I also really like some of the design features and construction of my AZ hives.

- ► **Hive body:** I love having 4 chambers.
  10 frames per chamber. All the space I need. 3 chambers would be adequate.
- ► **Bee entrance:** The bottom entrance has a full length opening with a mouse guard covering it. Sliding metal closes off opening as needed.
- ► **Bottom space:** screen and board or drawer.
  - • Under last divider board and at bee entrance level have extra space (hang out space). I have enough for an Oxalic Acid wand and easy cleaning of dead bees or junk. ¾ to an inch. I've never had any bur comb here.
  - • Screen is under this space. It keeps the bees in the hive. It allows for air flow, and a place for a dead mite to drop and some hive other dropping items (pollen, bee parts, wax from robbing, etc.).

Hive entrance 2020

- A board or drawer under the screen. Allows for a dead mite count and other droppings checking space. Also, it can control air flow. It gives easy access without bothering the bees. Located under hive outer door.

▶ **Inner hive doors:**
  - Screen on the inner door is great to allow looking at the hive all year long.
  - Solid board to cover the screen for treatment or just to close.
  - P-latches to close/lock doors in place. One on each side allows you to make tight. Or leave a space for a temporary Langstroth frame. Thanks Dana Schack.

P latch inner door 2020

Lift off hinge 2020

▶ **Metal spacers:**
  - Metal frame spacers work great. Two in front of hive and two on the inner door in the back. I also have them on the frame stand, frame spacer board and in nucs.

▶ **Feeders:**
  - While feeders are not part of the hive (extras) they are usually in my hives. Quart jar feeders that bees have access only from inside the hive (screens) help me monitor what extra food they are eating. The 3-jars make it so I can leave as much food as needed. The lids have small holes for bee's proboscis (tongue) to get the sugar water.

▶ Frame construction:

- Also, not part of hive but essential. I connect the sides different from original style AZ frames for better support. Top and bottom are the same concave on one side. Sides are just straight pieces. Nail through top/bottom to connect sides. A brad air gun works great for this. Nails larger than a brad may split the board. The extra strength is an advantage if propolis is holding frame while pulling it out. Cutting the concave side is on two pieces not all four sides of the frame. Thanks to my daughter Michelle Boyer, structural engineer.

▶ Langstroth Frames:

- I cannot recommend using Langstroth frames in AZ hives or recommend the hybrid Langstroth AZ hive. The AZ frames are designed to work with AZ hive. The Langstroth frame was designed to work with the Langstroth hive. The Langstroth frames collect too much propolis and bur comb at ends and on the bottom. This caused them to be extremely hard to remove. Tabs, extra width on top of end and the flat bottom are the issues.

Jar feeder 2017

▶ Hive Stands:

- I built a sturdy shelf for the hives to sit on. The shelf needs to hold a hive or hives full of honey. I built mine 20 inches off my floor. I store tubs under my hives. This was also a good working height for me. If you have stacked hives, you may not have much space from the floor.

AZ frame connect 2020

AZ Drawings Debra's Basic AZ Hive Drawings

AZ Bee Hive Body
(door side view)

Hive is 4 chambers tall with 10 frames in each chamber.

The outside board in measurements are not included only inside measurements. The outside measurement depends on the thickness of the wood you use.

<u>Hive/Chamber width</u> = 10 frames (1" each) = 10 inches. Bee space (3/8" each) on each side of frame for 11 spaces = 4 1/8 inch. For a total of **14 1/8 inches** inside measurements.

▸ A ¼ inch extra is nice on the outside sides to move frames. Bees will possibly add more comb to the frame next to that side though.

▸ For side boards (outside) of the hive, add 1" each side depending on the thickness of the wood.

<u>Chamber height</u> = top bee space 3/8", frame 9 1/2", frame support bar 3/8", 1/4" extra, separator board 1/2" (solid, separator, or queen excluder 11 1/2 x 18") = **total 11 inches**

▸ No top or bottom boards in measurements.

▸ ¼ inch extra is taken up with L bracket holding divider board, space above divide board so that it slides. Any left is good for top of frame space (makes moving them easier).

<u>Hive/Chamber depth</u>= Frames 18", two frame spacers (1/4" ea. One is on inner door) 1 ½", plus ¼" extra (Inside hive depth 18 ½"). Plus feeder 4", inner door ¾ wood , (open cell foam insulation space). **Total hive depth 23 ¼"**

▸ Does not include the exterior front or back outer door. Add insulation space as needed to front. I recommend solid wood to prevent pests from invading an insulation void.

<u>Total Hive height</u> = 4 chambers (44"), bottom bee space 3/4" = **total 44 3/4"**.

▸ Top and bottom board of hive add 1" outside. Or bottom assembly will add more.

▸ Add more above hive if you add insulation or in my case storage for divider boards.

Suggestion: make 1-2 chambers and then attach chambers to get the amount you wish.

AZ bee Hive (door side view)

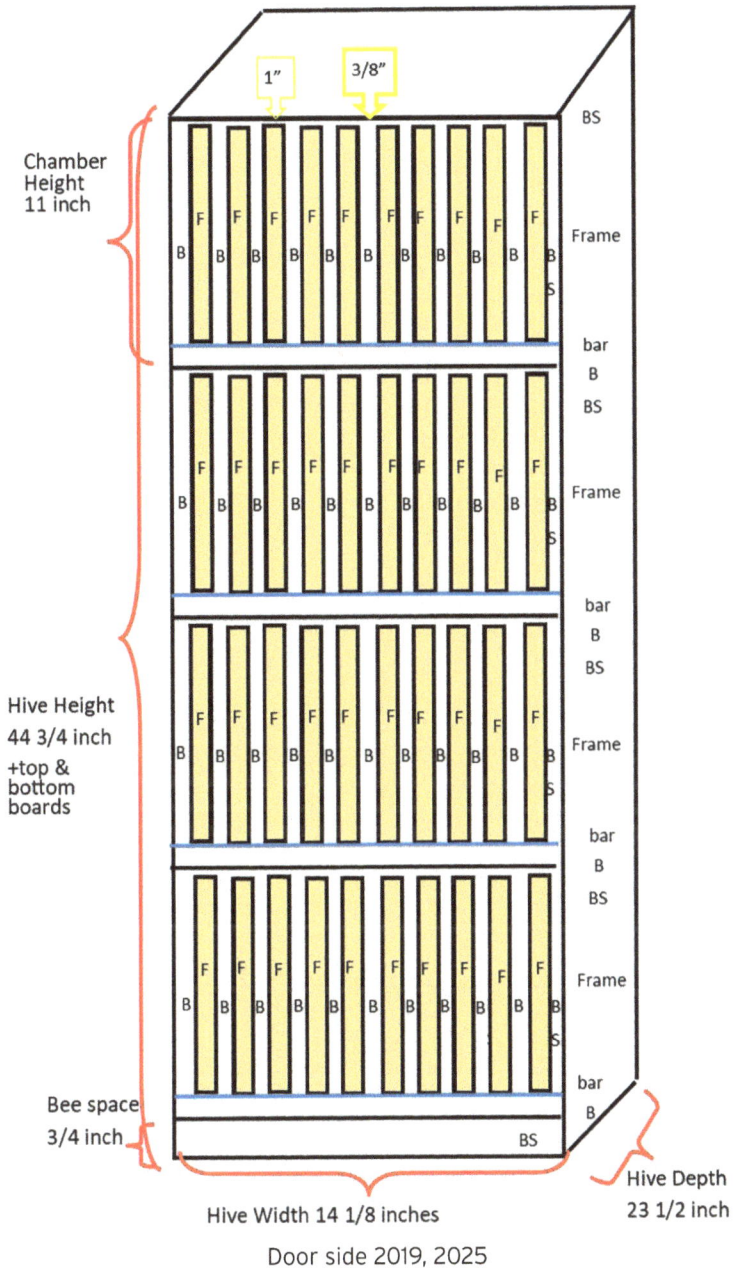

1" 3/8"

BS

Chamber
Height
11 inch

F F F F F F F F F F
B B B B B B B B B B
                          S

Frame

bar
B

BS

F F F F F F F F F F
B B B B B B B B B B
                          S

Frame

bar
B

BS

Hive Height
44 3/4 inch

+top &
bottom
boards

F F F F F F F F F F
B B B B B B B B B B
                          S

Frame

bar
B

BS

F F F F F F F F F F
B B B B B B B B B B
                          S

Frame

bar
B

Bee space
3/4 inch

BS

Hive Width 14 1/8 inches

Hive Depth
23 1/2 inch

Door side 2019, 2025

Drawing not to scale

Bee Space = 3 /8 inch = BS

Frame = 91/2 x1x1 inches = F

Separator Board = 1/2 x19x141/2 inches = B

(Add 3" handle & 1' to length 20")

Bar= 3/8" by about 16 inches (extra is for slots)= bar

145

AZ Bee Hive (side hive depth)

Do as much as you can before connecting sides. This is much easier than trying to gain access once sides are attached.

▸ Draw all items on side panels for placement.

▸ Attach bracket hardware for separator board on each chamber.

▸ Drill holes or slots for bars.

▸ Mark or drill slots for the inner door stand. Mark or drill at the same height as the bars. This is a place for a 2" board that holds the inner door.

Notes:

▸ If there is extra space at the top, it can be used for insulation or storing divider boards.

▸ If needed, add a block strip to front insulation space to nail front inner & outer board on.

Bar placement per chamber. Height (1 1/4", 12 3/4", 23 5/8", 35") and width -from front to back (5", 10", 15"). Bars are evenly spaced with one in center.

Bars are 3/8 plexiglass electric fence poles (4' cut 3 per). Or bar can be a metal bar that is threaded at the ends and screw in for better support of sides.

Slots allow for bar removal. Make a width to hold a 3/8" bar. Cut straight up from bar spot for about 5 inches. Cut into sides about ½ inch (need thick side boards for this). If skilled you can cut slot at angle for easier removal.

Bracket to hold separator board. Metal L bracket 18" long each side and about 14 1/8" for front. Place facing down for screws. This will be on both sides and the inside front (bee side) of the hive.

Hive/Chamber depth= Frames 18", two frame spacers (1/4" ea. One is on inner door) ½", plus ¼" extra (Inside hive depth 18 ½")   Plus feeder 4", inner door ¾ wood , (open cell foam insulation space) . **Total hive depth 23 ¼"**
Does not include exterior front or back outer door. Add insulation space as needed to front. I recommend solid wood.

Drawing was done with insulation space. Need to add the 1" insulation and inner and outer board. Measurements on drawing reflect this.

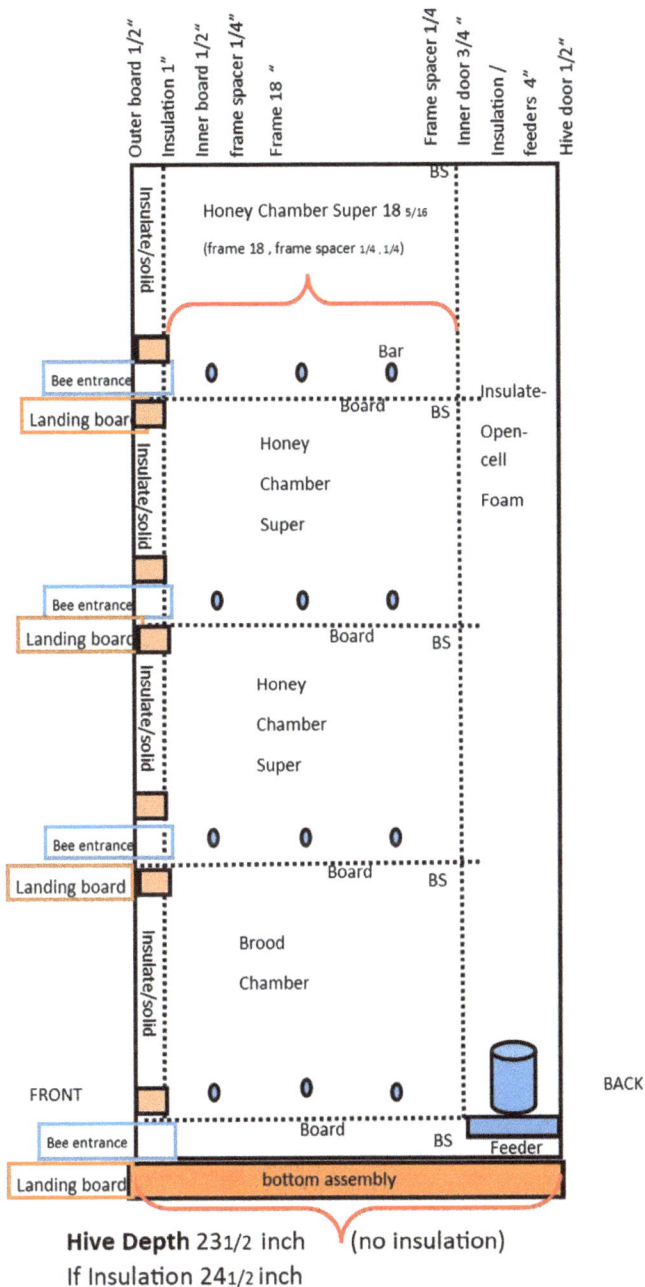

Side view 2019, 2025

Column labels (top, vertical text):
- Outer board 1/2"
- Insulation 1"
- Inner board 1/2"
- frame spacer 1/4"
- Frame 18 "
- Frame spacer 1/4
- Inner door 3/4 "
- Insulation / feeders 4"
- Hive door 1/2"

Insulate/solid

Honey Chamber Super 18 5/16

(frame 18 , frame spacer 1/4 , 1/4)

BS

Bar

Board   BS

Insulate-
Open-
cell
Foam

Bee entrance
Landing board

Honey
Chamber
Super

Board   BS

Bee entrance
Landing board

Honey
Chamber
Super

Board   BS

Bee entrance
Landing board

Brood
Chamber

FRONT   BACK

Bee entrance

Board   BS

Feeder

Landing board   bottom assembly

Hive Depth 231/2 inch   (no insulation)
If Insulation 241/2 inch

Drawing not to scale

Bee Space = 3 /8 inch = BS
Frame = 91/2 x1x1 inches = F
Separator Board = 1/2 x19x141/2 inches = B
(Add 3'' handle & 1' to length 20'')

Bar= 3/8" by about 16 inches (extra is for slots)= bar

147

AZ Bee Hive side panel (hive depth)
Two chambers with slots

- Follow directions previous directions: Draw on board, attach, drill,
- Two, or any number of chambers hive is similar, just different total size.
- The chambers are the same.
- No insulation space on this drawing. Add thicker wood to the front total size change with no insulation

  (-1" insulation and inner panel).

Adjust your hive to the size you want.

- Frame number, size and bee space dictate the size of the chamber.

Follow directions above: Draw, attach, drill,

▶ Draw all items on inner board.

▶ Attach frame spaces on inner board.

▶ Attach blocks, if needed, on the other side of inner board.

▶ Cut bee entrance on inner board + block.

▶ Connect to sides.

▶ Add insulation. Or use two boards together instead for insulated area. (fill area with solid wood).

▶ Front board attached on outside of hive.

▶ Cut bee entrance, add mouse guard and landing boards to front of hive.

Frame spacer –6/8" spacing. Each point and between each point. Flat bar at 90˚ angle to connect. Length varies. I used 10 frame.

Placement of frame spacers on backside of inner wall. Two per chamber. Bottom up (4 1/2 + 9) per chamber (also two are placed on inner door)

Bee doors. Each chamber has entrance and may have landing board with hinges & hook for closing. Bottom ones sized for mouse guard. Located at bee space just above separator board, & at the bottom of the hive. Top 3" are 6" x 3/8", bottom 10" x 1". Top landing boards use hinges to open or close or disk (dial) opening.

If using insulation on front. Blocking Bee doors. Place block (13" 1x 4 board) over space where bee entrance is to keep bees from insulation. Cut through block and boards.

In my wooded location I find carpenter ants may get into any foam insulation even with a good seal.

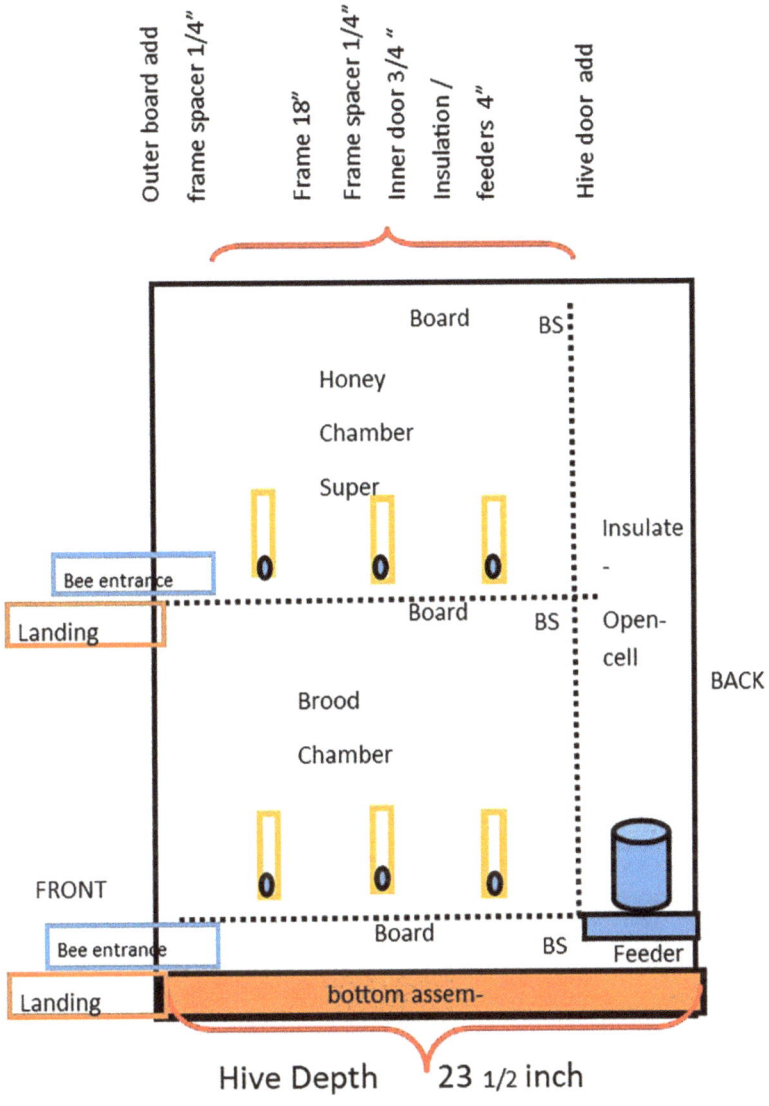

Outer board add
frame spacer 1/4"

Frame 18"

Frame spacer 1/4"

Inner door 3/4 "

Insulation / feeders 4"

Hive door add

Board — BS

Honey

Chamber

Super

Insulate -

Bee entrance

Landing — Board — BS — Open-cell

Brood

Chamber

BACK

FRONT

Bee entrance — Board — BS — Feeder

Landing — bottom assem-

Hive Depth — 23 1/2 inch

Two chamber side 2019, 2025

Drawing not to scale

Frame spacer 2019

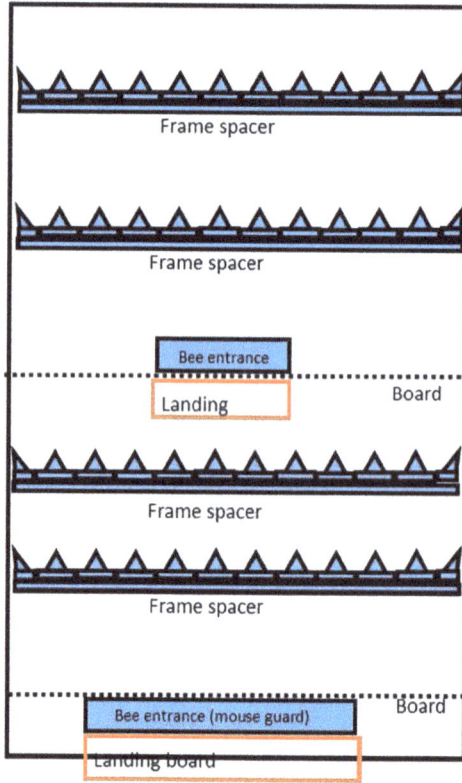

Frame spacer

Frame spacer

Bee entrance

Landing   Board

Frame spacer

Frame spacer

Bee entrance (mouse guard)   Board

Landing board

Two chamber inside front 2019, 2025

Drawing not to scale

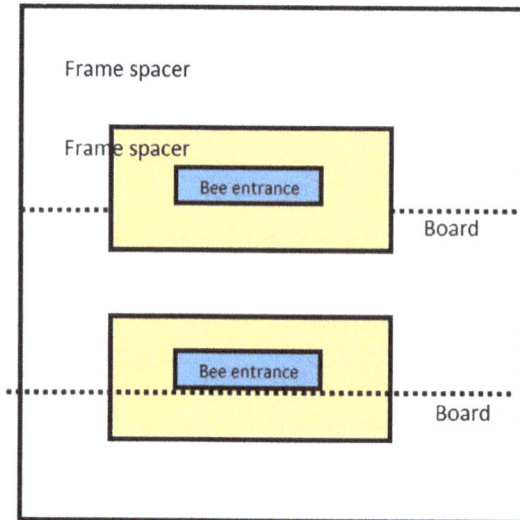

Frame spacer

Frame spacer

Bee entrance   Board

Bee entrance   Board

Two chamber outside front 2019, 2025

Drawing not to scale

Bottom of AZ Bee Hive

Drawings not to scale.

Bottom board –top framed 1/8' screen front solid back open. (screen with space for board under to close off or add mite controls). Hole for screen no more than 3" in from sides of chamber.

Varroa mite drop space needs to run the depth of the hive to be able to pull out drop board/drawer. This is also a good place to help control hive beetles

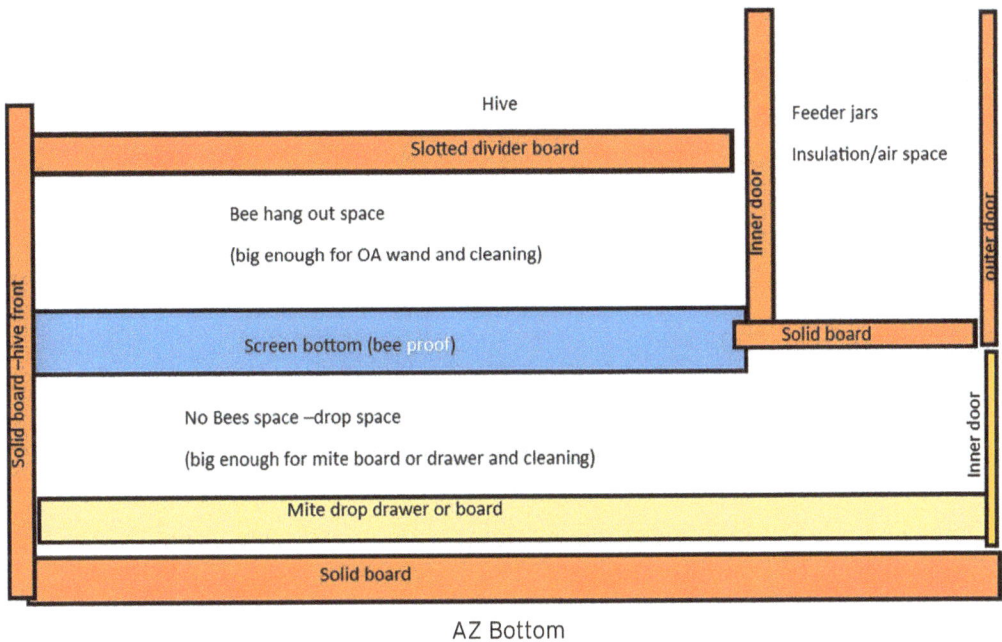

Hive

Feeder jars

Slotted divider board

Insulation/air space

Inner door

outer door

Bee hang out space

(big enough for OA wand and cleaning)

Solid board –hive front

Screen bottom (bee proof)

Solid board

No Bees space –drop space

(big enough for mite board or drawer and cleaning)

Inner door

Mite drop drawer or board

Solid board

AZ Bottom

Drawing not to scale

AZ Bee Hive Door

Drawings not to scale.

See photos chapter 3 The Hive (AZ hive) and 6 Designs for the AZ (hive construction)

Inner doors —plexiglass or screen 14 3/4 by 10 3/8 by 3/4 at each chamber.

The Inner door slides into hive holding frames in place.

▶ The center can have a screen, plexiglass and/or removable (second frame).

- ▶ Handle & clips or bar to hold door closed.
- ▶ Frame spacers attached to bottom/top inside of inner door.
- ▶ Do not make too tight (wood can swell and shrink in some climates).
  I've stapled a folded strip of outdoor fabric on the outside edge (top & sides) to close any bee gaps (escapes). The fabric moves as needed.
- ▶ Make a solid board to cover screen for winter and treating.
- ▶ Add small hole for bottom for feeder. Make a plug (a piece of foam works)

On hive add 2 inch board for the inner door to sit on. This is at the exact same height as the bars. The frames can then slide out onto this board. But not to interfere with the divider board removal.

Air vent add vents to top of each chamber on outer door with 1/8" screen. Have a hinged cover to open or close vent as needed.

Inner door back 2024

Outer door (Back) Attached door on outside of hive with lift off hinges and handle. Door should cover at least two chambers.

Door can have an added chalkboard paint, or cork board, or other way for notes on the door. (On inside or out)

The bottom of the door should also cover bee hang out space.

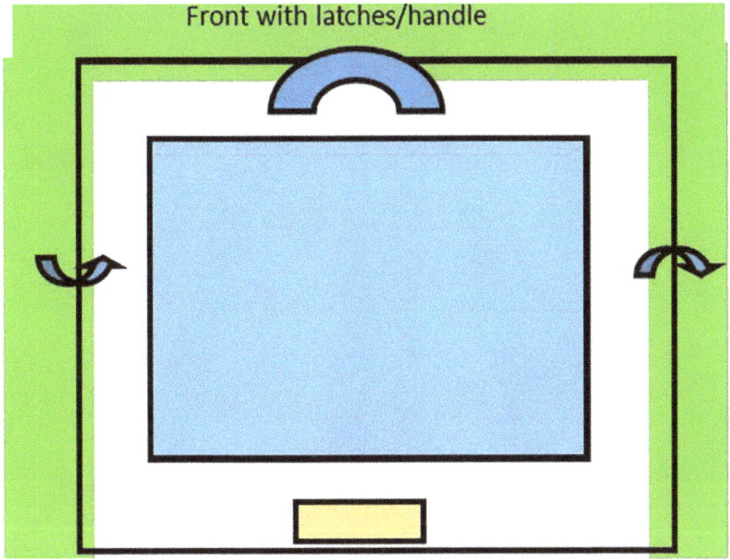

Front with latches/handle

Bee feeder hole

Inner door front 2024

Drawing not to scale

Back with Frame spacers
Inner door bee side 2020 photo

Inner door front 2020

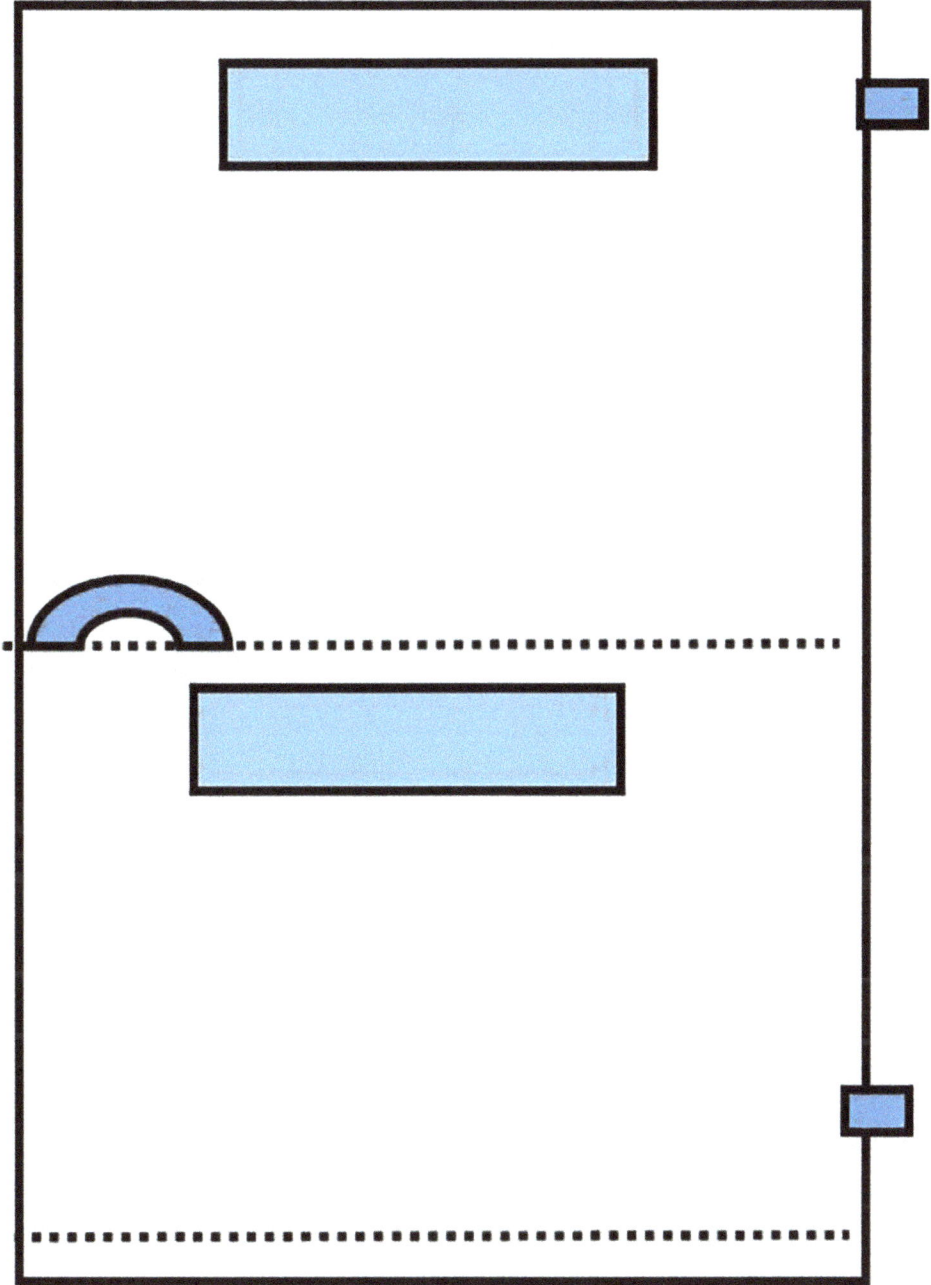

Outer door 2024

AZ Boyer Hive Frame

American AZ Hive (Langstroth size frame)

Frame fits a Langstroth deep plastic foundation. Drawings not to scale. Frame measured from outside edges. Drawings not to scale. See photos chapter 3 The Hive (AZ hive) and 6 Designs for the AZ (hive construction)

Cut 2 side pieces for frame 1" x 1/2" x 8" no routing needed.

▸ Cut two pieces for top & bottom of frame 1" width x 3/4" x 18". The 1-inch width gets cut into both sides. The tools available will decide the cut depth and width.

   a. Top of board. The top is concave the entire length. About 1/4" deep cut and about 3/4" wide. It is easier to route longer boards first then cut to size for the frame.
   b. Bottom (inside to hold foundation) of board has groove the length of the other side of these boards about 1/16 " x 1/4. Easy to cut longer boards first then cut to size for frame.

▸ Connect to make frame. Can nail, screw or dovetail and glue. If using plastic foundation put in before you connect the top of the frame.

Frame Notes:

Read articles in other chapters "Frames", "Lang to AZ Frames", and "AZ Hive Construction" for more information.

This frame design is different than other AZ frames in the way the sides are connected. Design by Michelle Boyer (structural engineer). "AZ Boyer Frame"

The frame is the key to the size of the hive chambers. Design it first.
Then decide the number of frames in each chamber. This gets you the base for the whole hive.

- Use good wood for the frames. Bees create moisture (warping) and pressure is put on frames for removal when the bees do a bunch of burr comb or propolis on it.
- Route long **concave** pieces. They can be cut to size later. Must for the bottom of the frame touching the bar. Possibly it is not needed at the top of the frame. But cannot flip the frame over if not concave.
- Glue the joint if you can. Dovetail joint would be the best, but do what you can. I just nailed them, and they worked fine.
- Make a channel to insert **plastic foundation** if you plan on using it. Cut on other side of concave piece.

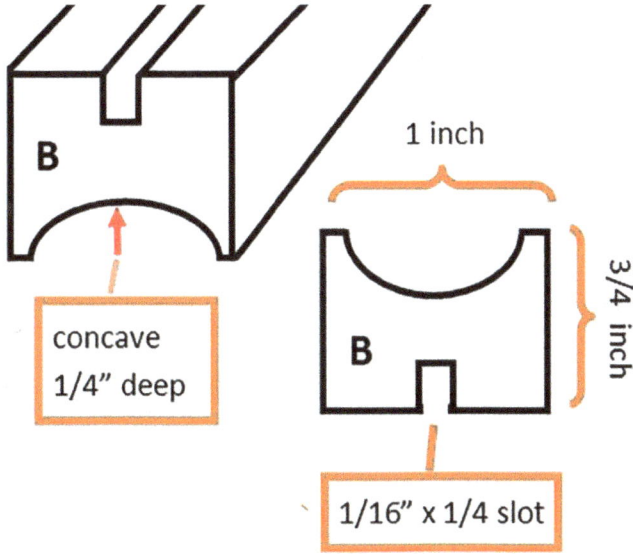

concave
1/4" deep

1 inch

3/4 inch

B

B

1/16" x 1/4 slot

AZ frame ends 2019

- Make extras for future replacing of old frames. I would do 10 for each brood chamber. Do this since it is not easy to have a source for new frames.
- Do not put together all the frames that you are not using—they store better in stack.
- Leave some of the top bars off if you are getting Nucs. It is easier to slide foundation in and then seal.

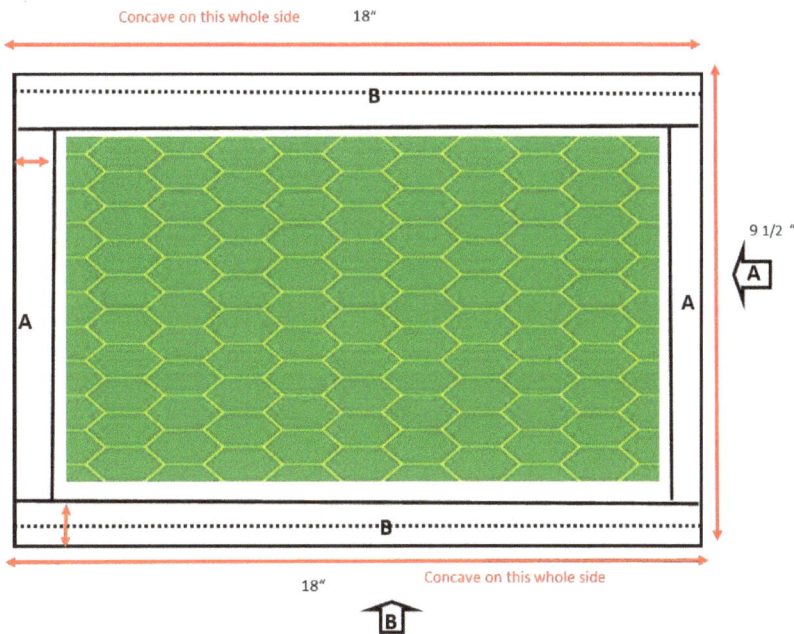

Concave on this whole side   18"

B

A

A

9 1/2 "

A

B

18"   Concave on this whole side

B

AZ frame drawing 2019

Pictured is the Slovenian frame construction of sides to top & bottom. Note that the sides are longer and connect to ends of top and bottom of the frame. If doing this method you need to route the sides top and bottom concave.

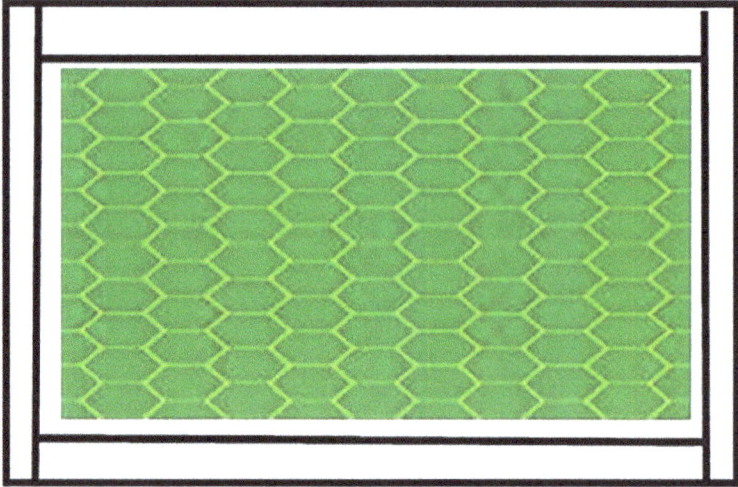

Slovenian AZ frame drawing 2019

AZ Bee Hive divider boards

Drawings not to scale. See photos chapter 3 The Hive (AZ hive accessories)

Divider Board—14 ½" x 20" You can do Solid, Slotted, or Queen excluder. These are to separate the chambers. Add a handle to pull in and out (add 3" for handle)

Slotted divider board Bees can pass through. Frame with spaces between boards placed to leave bee space. I run the slots the long way. You can also do round holes for bees to pass (currently testing hole design, size and number of holes could vary)

Queen excluder divider board Frame with metal queen separator inside. I used a Langstroth 10 frame queen excluder with board around the edge.

Semi-Solid divider board. Hole in end for bees to pass through. This can help hold heat in the winter. Also it allows bees to rob their own honey frames in the upper chamber.

Solid divider board 2024

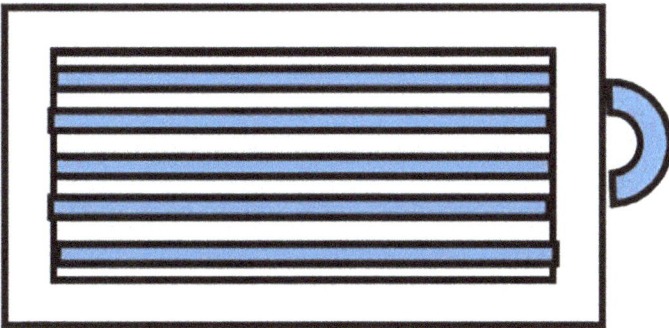
Slotted divider board 2024

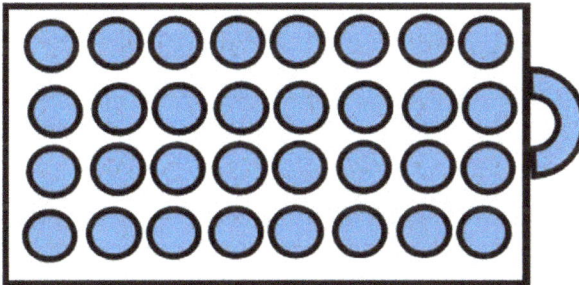
Holes divider board 2024

Slots wrong direction 2024

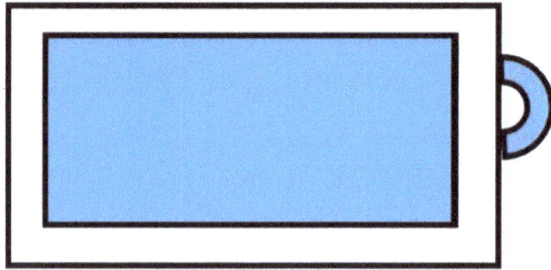

Queen excluder divider board 2024

Semi-solid divider board 2024

AZ Bee Hive Feeders

Drawings not to scale.

Sugar Water feeder jar holder

On the inside of the **Bottom** of hive chamber, I have a place for a **feeder**. Use the whole space along the back to accommodate 3 jars. Bees enter from the hole in bottom of inner door to access the feeder. The bees can feed through the screen and jar lid.

Feeder design

▶ Top 4" wide x 14 3/4" long piece.
  - Cut 3 holes for jars to fid down in. Check size of your jars with lid.
  - Covered bottom side of hole with bee proof (1/8") screen to keep bees in when changing jars.
  - Nail this on top of sides.
▶ Front side cut 143/4" long x ¾" tall x ½ wide.
  Height and width can change.
▶ Sides cut two 3½" long x ¾" tall x ½ wide.
▶ Leave the side next to the hive open but close the other three sides (front and sides) to contain bees.
  No bottom.

Jar feeder photo 2024

Jar feeder photo 2020

Use quart jars and lids. Lid should have tiny holes just big enough for bee to access. Jar must sit tight so bee can reach sugar water.

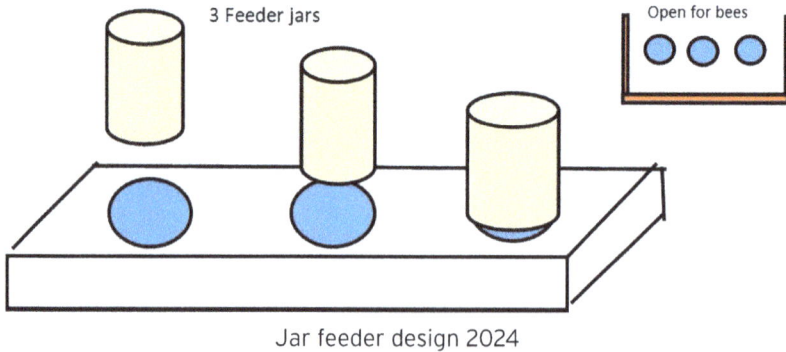

Jar feeder design 2024

AZ Bee Hive Table

Drawings not to scale. See photos chapter 3 The Hive (AZ hive accessories)
▸ Legs or other way to have the hive hold it solid while you work need to slide into the hive. Measure where you wish to place them and cut accordingly. I slide into the bottom hang out space under the last divider board. Working the AZ Hive (inspections & working AZ hive)

Cut board 18 to 20 inches by about 14 inches. (for table space)
Frame is a 1 x 1/2 inch board. Put frame on top (to keep bees and stuff from falling off) and bottom (to fit under divider board and bars/rods and keep bees in hive).

Bottom: Cut two legs 22 to 25 inches long and nail to long sides of board. (Frame legs need to be about 4 to 5 inches longer to slide under the bar inside of hive) Cut top /bottom cross pieces (about 14 inches depending on board). Nail to legs and board. At the top and bottom. Put the frame on top sides (to keep bees and stuff from falling) . Cut about 18 inches long. Nail on.

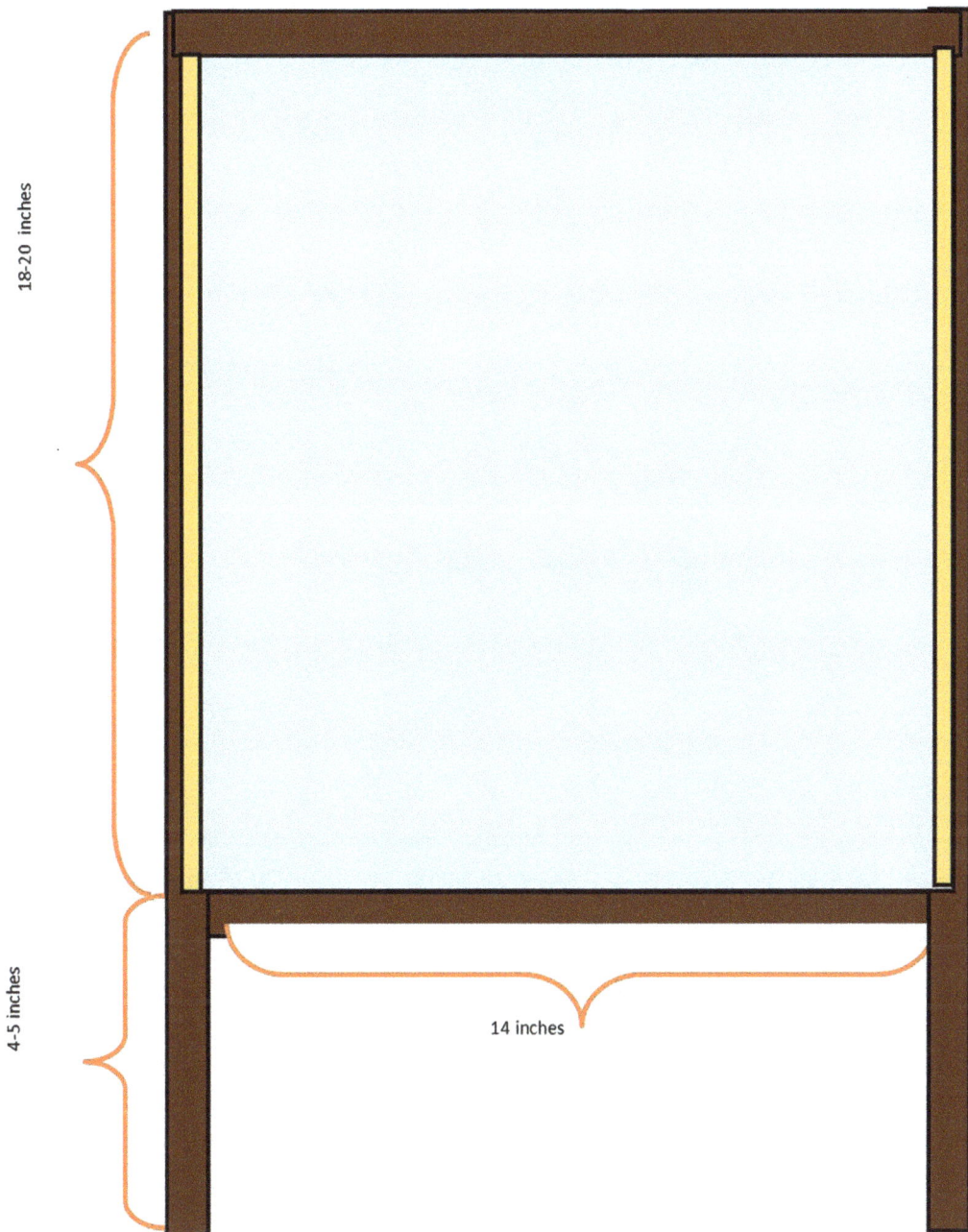

18-20 inches

4-5 inches

14 inches

AZ Hive table 2024

AZ Bee Hive Quilt Box

Drawings not to scale. See photos chapter 3 The Hive (AZ hive accessories)
I put two 4x4 posts about 16 inches long on top of bars/rods next to the sides of chamber above the colony. I set the quilt box on top of them. This creates a good space for winter emergency food under the quilt box.

The Quilt box is simply a box. The sides are about 5 inches tall, 16 inches deep and 14 inches wide. Just enough to fill most of the space in the chamber left after the 4x4 posts get in. I staple a small mesh metal screen on the bottom. You can add burlap bag and wood chips for moisture in the box.

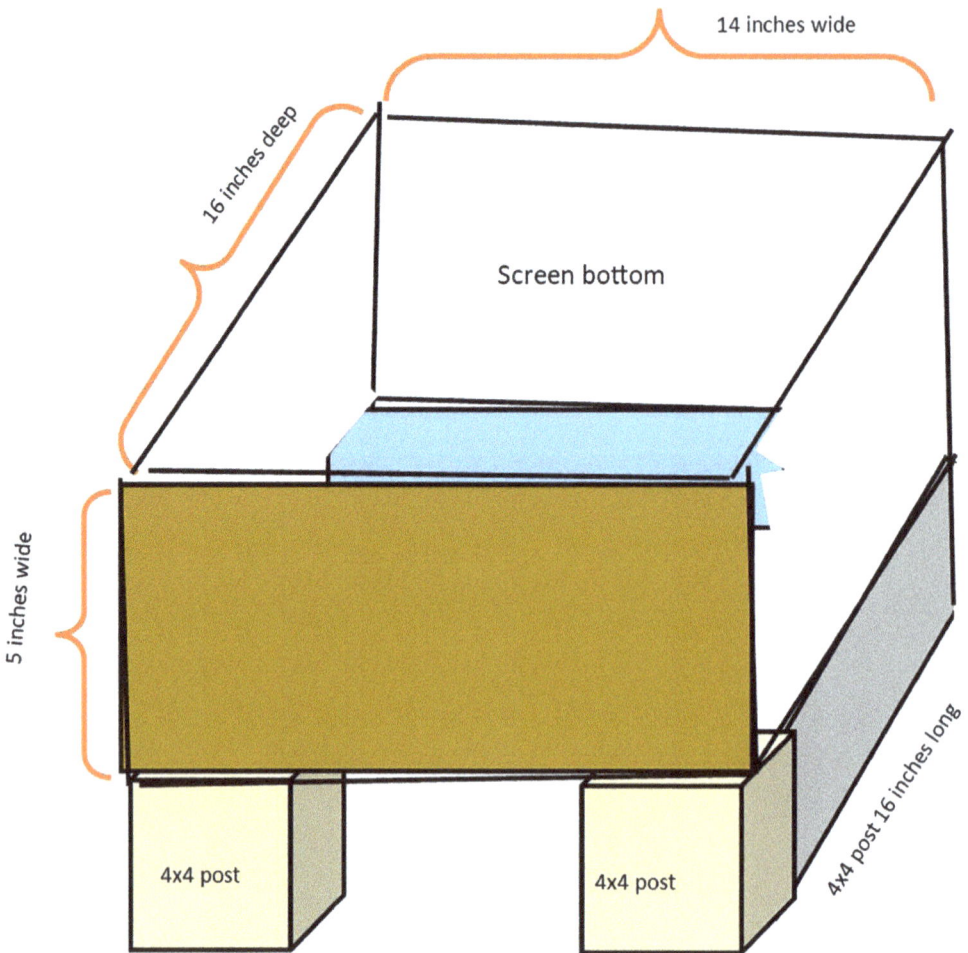

Quilt box in AZ 2024

AZ Bee Hive Robber Screens

Drawings not to scale. See photos chapter 3 The Hive (AZ hive accessories).

My robber screens vary in size with each hive entrance. The design is the same. They are about 8 inches tall, 10 inches wide and 2 inches deep. The front is covered with fine 1/8" screen. The top front half is covered with solid board. Working the AZ Hive (issues & solutions). Place a board at the top front to keep robbers only seeing the regular bee entrance at bottom of robber screen. Wasps go for that visible entrance not the hidden one at the top for the bees. The board also allows hive bees a place to be safe, but not seen. The sides are solid wood. A hole is on the top back to one side of the robber screen for the bees' hive entrance. It has a flap to close. Which is a small piece of wood with screw on back corner. The robber screen screws into the hive over the entrance.

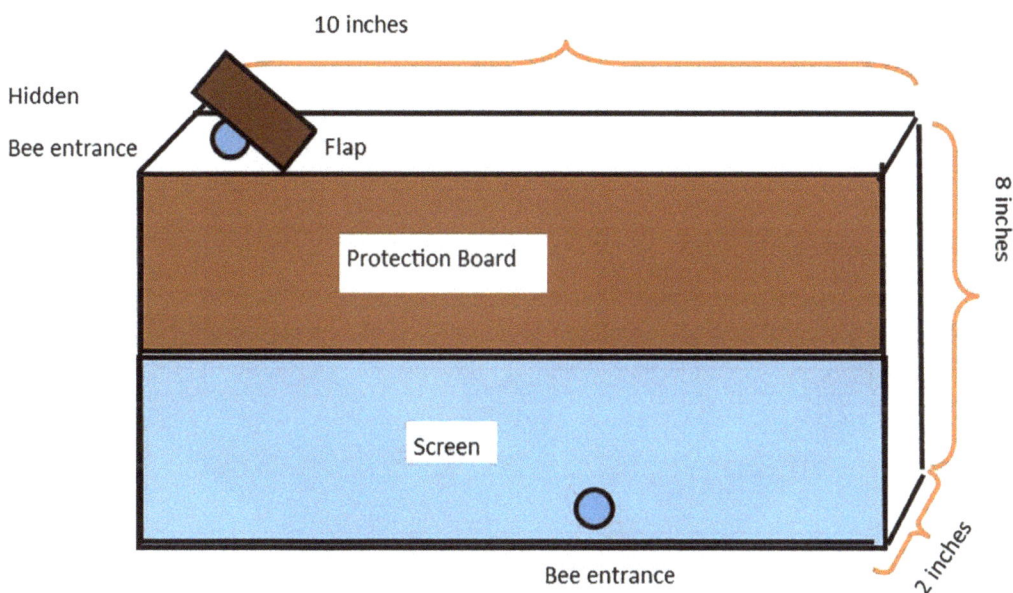

10 inches

Hidden

Bee entrance

Flap

Protection Board

8 inches

Screen

Bee entrance

2 inches

Robber Screen 2024

AZ Frame Stand/Holder

Drawings not to scale. See photos chapter 3 The Hive (AZ hive accessories).

Make your frame stand/holder for at least 10 frames. You can adjust to make sides at an angle (A) or one side flat (B). These are two different ways or styles you can make the stand A or B. I used excess wood that I had. Working the AZ Hive (inspections & working AZ hive) **Style A:** Set at an angle and can fold closed

▸ Small board - 10 by 16 inches -1 frame spacer
▸ Large board - 20 by 18-2 frame spacers.
▸ Screw frame spacers on each board 5 inches from the ends.
▸ Cut four legs for the sides (use 1 x 1, inch boards)
  • cut two at 25 inches long (this gives 5-inch legs on the larger board)
  • cut two at 15 inches long (this gives 5-inch legs on the smaller board)
▸ Connect legs to top of the sides of the boards leaving a 5-inch leg on each of the four sides
  • The smaller board & legs fit inside the larger one to connect.
▸ Connect the two boards with legs at the top of the five-inch legs by the boards.
  • Use a bolt to connect as this allows it to fold open and closed.
  • You need a wing nuts or other way to lock open when in use.

6.37 Folding frame stand A

Style B: Sets flat on one side

▸ Cut two boards:
  - Small side- 10 by 18 inches,
  - with 1 frame spacer
  - Large side- 20 by 18 inches,
  - with 2 frame spacers.

▸ Screw frame spacers on each board

▸ 5 inches from ends.

▸ Cut four legs/sides (use 1 x 1, inch boards)
  - cut two at 15 inches long (gives 5-inch legs on smaller board))
  - cut two at 5 inches long (gives 5-inch legs on larger board)

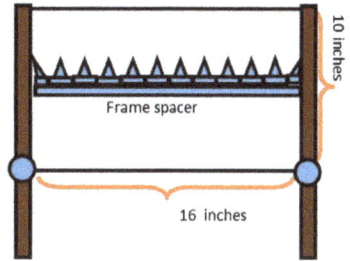

Small side frame stand A drawing

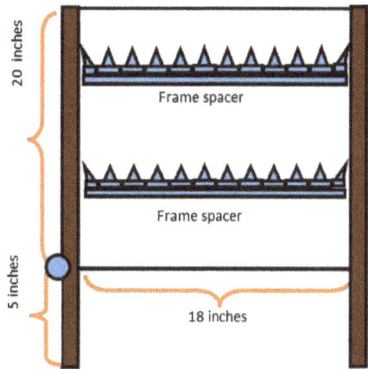

Large side frame stand A drawing

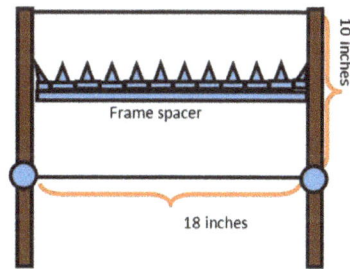

Small board B frame stand

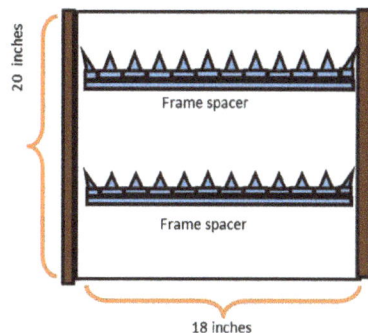

Flat frame stand B

Large board B frame stand

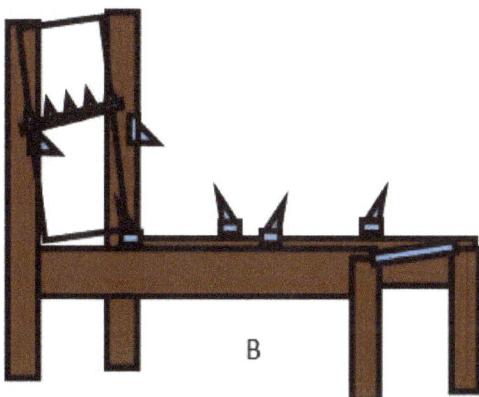

▸ Connect short board the same as with style A leaving five-inch-long legs.
▸ Connect the larger board to the smaller one on the side where boards can meet. (see drawing)
▸ Add 5-inch legs to the underside of the larger board.

AZ Nuc - 5 –frame

Drawings are not to scale.

All inside measurements are shown.

No outside boards added in measurements or Varroa mite drawer assembly.

These dimensions will vary depending on the thickness of boards used.

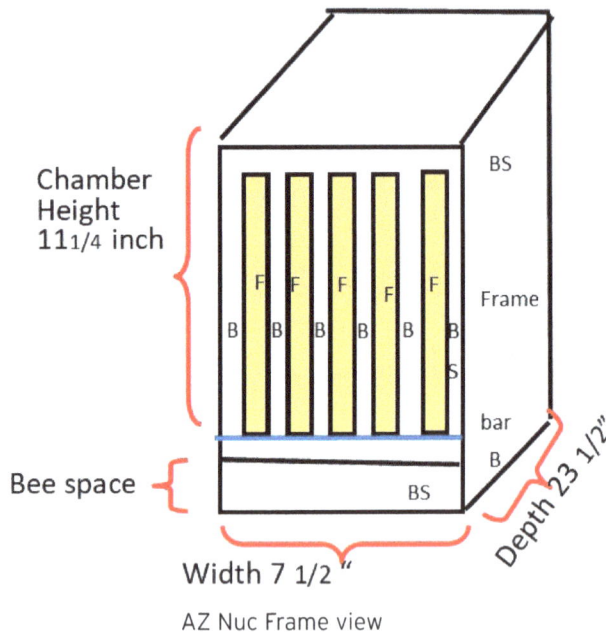

AZ Nuc Frame view

A Nuc is just a mini hive. Often the same design as regular hive.

See "AZ Hive Body" drawings and directions.

▸ Outside boards are ½ inch thick, or more if desired.
▸ Draw all items on side panels for placement.
▸ Attach hardware for divider board.
   • L brackets. Place about ½" below bars for divider boards to rest on.
▸ Drill holes for bars or add slots for removal.
▸ Divider board = ½ x 7 ¼ x 20 if bottom assembly
▸ Add handles to the sides to carry.
▸ Add latches on door to keep closed in transport.
▸ Screw solid bottom board for easy cleaning removal.

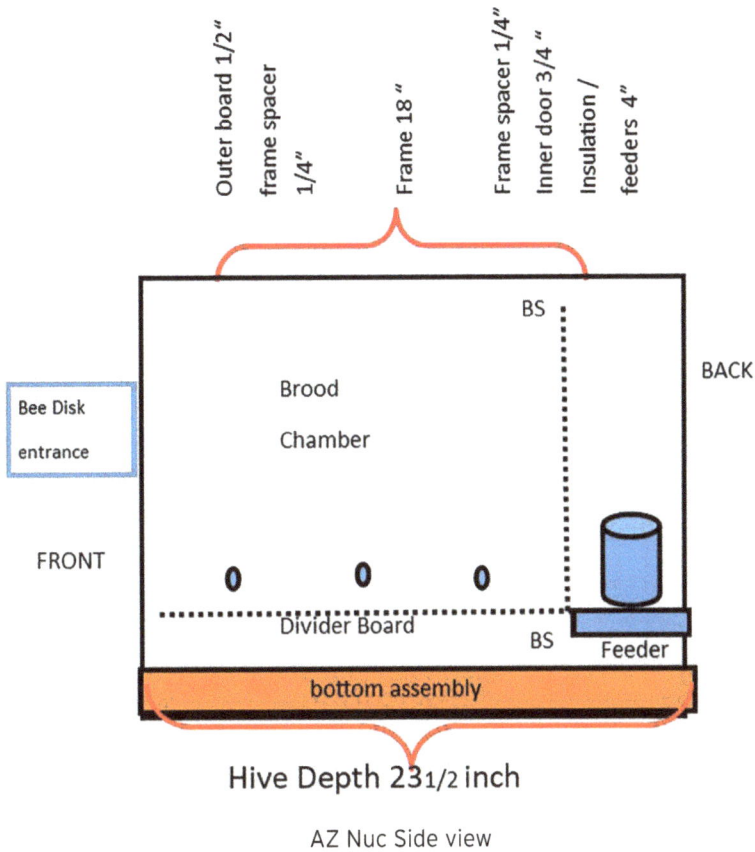

Hive Depth 23 1/2 inch

AZ Nuc Side view

Drawing not to scale

Bee Space =3/8 inch = BS

Frame =9 1/2" x 18 1/16 = F

Separator Board = 1/2" = B

Nuc height = bee space 3/8", frame 9 1/2", frame support bar 3/8", divider board 1/2", 1/2" extra = **total 11 ¼ inches**

Bottom of hive can be solid or add mite assembly)
▸ Mite assembly *if you wish*
▸ *(add to above measurements)*
   • add 1/8 inch screen to bottom
   • add about an extra inch at bottom for drawer or mite board. 12 ¾ inches tall

Nuc width = 5 frames (1" each) = 5 inches. Bee space (3/8" each) on each side of frame for 4 spaces = 1 1/2 inch. Space on sides ½" each for 1 inch. For a total of **7 1/2 inches inside** chamber.

<u>Nuc depth</u>= Frames 18", two frame spacers
(1/4" ea. One is on inner door) ½", plus ¼" extra (Inside hive depth 18 ¾")
Plus feeder 4", inner door ¾ wood , (open cell foam insulation space). **Total hive depth 23 ½"**

▸ Cut two side boards for outside 25 ½ by 11 ¾ inches
  • If mite assembly 25 ½ by 12 ¾ inches
▸ Cut solid bottom and top board   26 ½ by 8 ½ inches (overlaps all sides)
▸ Cut one board for back 8 ½ inches by 11 ¾ inches
  • If mite assembly 8 ½ by 12 ¾ inches
▸ Cut one front door board 8 ½ by 11 ¾ inches
  • Add lift off hinges, door lock and handle
  • If mite assembly 8 ½ by 12 ¾ inches
▸ Inner door frame 11 ¾ by 7 ½ inches. If too tight remove fraction from top and sides.
▸ See "AZ Hive Door" drawings
  • Attach 1/8 screen & screw on two
    frame spacers for 5 frames each (about 1 inch off top and bottom)
  • Add handle and P locks and feeder hole at bottom of frame
▸ Two frame spacers for 5 frames each. Add two to back of hive at about 3 and 7 inches down from top.
▸ Three bars. place front to back on sides (5", 10", 15"). Bars are evenly spaced with one in center. See "AZ hive Body" drawings
▸ Single hole feeder. 4 x 7 x ¾ see "AZ Hive Feeder" drawing

AZ Nuc-DS 5-frame

Design by Dana Schack
Drawing by Debra Langley-Boyer
Drawings are not to scale, these are inside dimensions.
See photos chapter 3 The Hive (AZ hive accessories).
<u>Nuc box, top feeder, extra space box and lid.</u>

▸ A solid center board is used to divide into two 2-frame hives. 25.5 x 7.5 (check fit, add outdoor cloth if needed)
▸ Each side has disk entrance and top feeder hole. For two colonies or queens.
▸ Add handles on the disk sides to carry.
▸ Add four latches/locks on the long sides of Nuc box and the top feeder to keep closed in transport.

- ▶ Add a board for top lid with overhang for rain. Different designs are possible. Can do the frame on under side for better closing.

Frame spacers:
- ▶ two 5 frame spacers on the bottom of the **bottom box.** (put about 5' from either end – see drawing).
- ▶ Four 5 frame spacers. Two for each box. Place about 5" down on each of the 7 ½" sides.

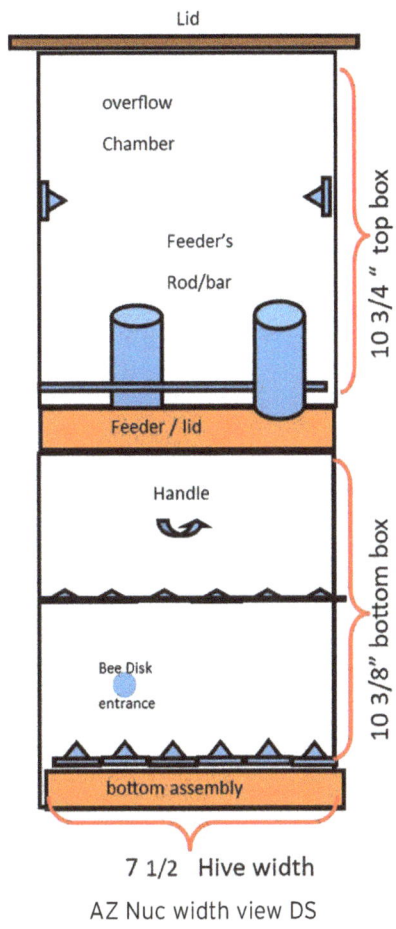

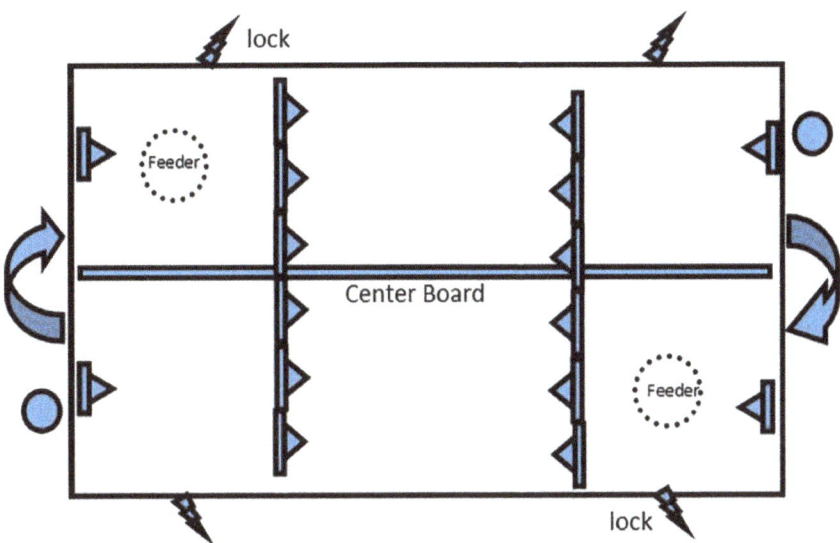

Lid

overflow

Chamber

Feeder's

Rod/bar

Feeder / lid

10 3/4 " top box

Handle

Bee Disk
entrance

bottom assembly

10 3/8" bottom box

7 1/2 Hive width

AZ Nuc width view DS

lock

Feeder

Center Board

Feeder

lock

AZ Nuc down view DS

171

Top Box add two 3/8" bars about ½" up from bottom of box. Place about 6" in from ends. (need to avoid place where feeder jar sits)

Feeder/lid assembly— Cut 1 inch board by 8 ½" x 19 ¾ inches. Cut holes for jar feeders at each end on opposite sides. Cut about 2" from ends. Add bee screen on bee side to keep the bees in.

Boxes use ½ inch boards for sides, bottom and top. Boxes inside measurements are shown.

Bottom Box

Cut two long sides 18 ¾" x 10 3/8"   Cut two short sides 8 ½" x 10 3/8"

Top Box

Cut two long sides 18 ¾ x 10 ¾"   Cut two short sides 8 ½" x 10 ¾"

Cut one bottom board 8 ½" x 19 ¾"   Cut one lid 11" x 22"

<u>Nuc bottom box height</u> = bee space 3/8", frame 9 1/2", frame support bar 3/8", divider board 1/2", 1/2" extra = total **10 3/8 inches**

<u>Nuc top box height</u> = bee space 3/8", frame 9 1/2", frame support bar 3/8", divider board 1/2", 1/2" extra = total **10 3/4 inches**

<u>Nuc width</u> = 5 frames (1" each) = 5 inches. Bee space (3/8" each) on each side of frame for 4 spaces = 1 1/2 inch. Space on sides ½" each for 1 inch.
For a total of **7 1/2 inches inside** chamber.

<u>Nuc depth</u>= Frames 18", two frame spacers (¼" ea.) ½", plus ¼" extra
Total hive depth 18 ¾"

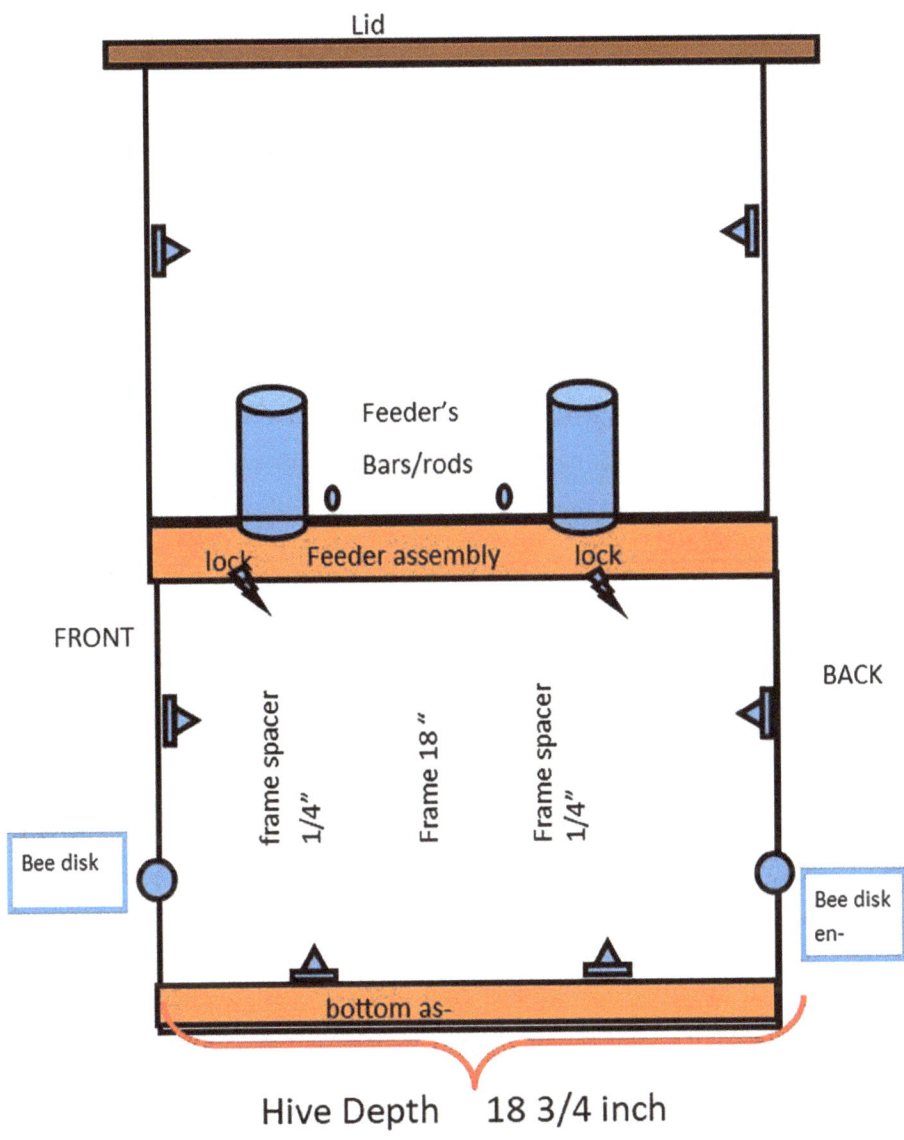

Lid

Feeder's

Bars/rods

lock   Feeder assembly   lock

FRONT

BACK

frame spacer 1/4"

Frame 18"

Frame spacer 1/4"

Bee disk

Bee disk en-

bottom as-

Hive Depth   18 3/4 inch

AZ Nuc down view DS

Reference & Credits

Credits:

Photos and drawings are by Debra Langley-Boyer unless noted. Some of me were taken by unknown beekeepers sent after attending an event at my house. Photo credits below are listed with chapter number and with photo. Thank you to photographers, those photographed and all for sharing. Some wished to only use their first name.

Photos owner or copyright or taken by:

Michelle Boyer (Acknowledgements), Brian Drebber (Dedication), Sherry Purkett (Preface, 3, 4, 5), Beekeeping museum in Radovljica, Slovenia (1), Gene Kritsky (1), Barbara (1), Simon Dollnsek (1), Vesna Sűssinger (1), Dana Schack (2)

People in photos:

Drebber Brian (Dedication); Ellen (1, 4); Debby (1); Marjeta Fupančič (1); Nedra (1); Barbara (1); Nika Pengal (1); Sűssinger, Aleš (1); George (2, 3, 4), Bryan (2); Shaari (2); Dana Schack (2, 5)

Design credits

Schack, Dana. (6) AZ Nuc

Printed material - Books, magazines, Facebook & internet

- ▶ Božič, Janko, 2015, *AŽ Beekeeping with the Slovenian Hive*, Založba Mija
- ▶ Pengal, Nika & Petelin, Anže Gallus, 2023, *The Secretes of Miss Honeybee*, Pengal, Nika & Petelin, Anže Gallus
- ▶ Copi, T. Rene, 2020, *Pobarvajmo Panjske Koncnice*, T. Rene Copi SP
- ▶ Cufer, Andrejka, 2018 *A Beekeeper at Heart*, SVETJESVET, Fineart creativity, Andrejka Čufer
- ▶ Langley-Boyer, Debra, May 2020, "*The AZ Hive Journey*" pg91-93, Bee Culture

There is information on the **internet**. Be careful as there are quite a lot of places with incorrect information and people selling untested newer ideas. There are several YouTube videos about AZ hives. I see more information posted every year.

Some **Facebook** groups that are for AZ hives. Most have information in the files section.

▸ AZ Hivers
▸ AZ Hive Creators – a place to design and build AZ hives.
▸ Slovenian AZ Hive Advocacy (AZ Hives North America).
▸ AZ Hives – Pacific Northwest
▸ AZ Hives Northern California

AZ Hive Suppliers

I am not recommending or endorsing any business. I have only listed a few of the USA businesses, Europe has many more.

▸ AZ Frame Spacers

• Offers the metal frame spacers. Florida, USA
  AZHiveSpacers.com azframespacers.com

▸ AZ Hives North America

• Offers hives, accessories. Rhode Island, USA
  www.azhiveshorthamerica.com

▸ MaineBees

• Slovenian AZ hive, Slovenian & American frames, accessories. Maine, USA
  www.mainebees.com

Slovenian Beekeepers, Apiaries, other bee related places in Slovenia

Places I visited on my trip in 2023. Many can be found in photos (chapter, number in that chapter). They are in the order I saw them on my trip to Slovenia. Member of Slovenian Beekeeper Association (SBA), European Beekeepers Association (EBA), Certified National Beekeeper (CNB), Nationally Certified Apitherapist (NCA), Board member, in charge of or active with a bee related organization. (Active)

▸ Beekeeping museum in Radovljica, Slovenia Beekeeping Museum – Museums of the Municipality of Radovljica
▸ Beekeeping Education Centre of Gorenjska in Lesce, Slovenia.
  Špela Kalan, apiary manager
  • www.circg.si
▸ Anton Janša's beehouse located in Breznica, Slovenia
▸ Bee Toni Apiary. Rateče, Slovenia. Toni & Marjeta Fupančič.
  Beekeeping-TONI | Facebook or http://bee-toni.eu
  • Artist Alenka Peternel Hubert

- Gospodična Medična & apiary. Domžale, Slovenia. Nika Pengal, apitherapist, instructor & author of The Secrets of Miss Honeybee. Facebook Gospodična Medična or www.Gospodična-Medična.com SBA, EBA, CNB, NCA, Active
- Tigeli Apiary. Kolodvorska, Slovenia. Dragica & Joel Tigeli. www.cebelarski-muzej.si | Beekeeping Museum, Beekeeping Tigeli
- Čajnica Apiary . Suete Ana, Goricah, Slovenia. Danica & Jože Kolarič www.cajnica.si Teahouse | Kolarič herbs
- Hauzer Apiary & Museum. Cogetinci, Slovenia. Joseph & Majda Hauzer
- Olimije beehouse located at Olimije monastery. Tended by the Monks
- Zavod Čebela Apiary. Novo Mesto, Slovenia. Andreja & Tjaša Stankovič. https://zavod-cebela.si
- Celelarstvó.dolinsek. Novo Mesto, Slovenia. Simon Dollnšek
  - Facebook cebelarstvo.dolinsek
- Skocjanske Jame caves beehouse. Divača, Slovenia. www.park-skocjanske-jame.si
- Matija Komac's beehouse apiary. Bovec in Soca Valley, Slovenia.
- Tomislav Rene Čopi. Bovec area, Slovenia. Painter of Panels. Author of *Pobarvajmo panjske koncnice paintings*
- Slovenian Beekeepers Association. Lukovica, Slovenia. Https://en.czs.si
- Plečnik Beehouse. Castle, Ljubljana, Slovenia. Aleš Sűssinger beekeeper. SBA, EBA, Active

www.ingramcontent.com/pod-product-compliance
Lightning Source LLC
Chambersburg PA
CBHW041801280326
41926CB00103B/4662